Heisenberg-Kepler WaveFunctions At Quantum Speed Of Gravity $C3$

Dc = c / V

T = 2π x ([GM/c³] x Dc³)

→ $T^2 = [4\pi^2/GM] \times R^3$

→ $\Delta x \Delta p \geq \hbar/2$

R = [GM/c²] x Dc²

→ $V^2 = GM/R$

ε = (6π / Dc²) / (1-e²)

→ $\varepsilon = 24\pi^3[a^2/T^2c^2(1-e^2)]$

Published by Antoine Khai Nguyen

Copyright 2024 Antoine Nguyen

First Edition 1.1

Table of Contents

Introduction — 10

    Naming Conventions In This Book — 12

Discovery Of Orbital Quantizer Mechanism As Gravitational Link In Orbital Velocity-Radius-Period-Precession Determination For Moving Celestial Objects: — 13

    Orbital Quantizer In A Nutshell — 14

    Frequency And Wave As Other Fundamental Nature Of Orbital Quantizer — 17

Discovery Claim Of Orbital Period Quantizer Mechanism Behind Kepler's Cosmic Laws: — 19

    Speed-Of-Light-Based Orbital Period Quantizer Mechanism's Equations — 19

    Deep Correlation Between Orbital Period And Orbital Radius In Quantizer-Based Creation Mechanism — 20

    Emergent Values Of Gravitational Motion Energy Potential Quanta — 22

Discovery Claim Of Quantum Superposition And Wavefunction Via Orbital Period-Radius-Velocity Quantizer Mechanism Behind Kepler's Cosmic Laws: — 24

    Quantum Superposition Revelations From The Equation Of Orbital Period Quantizer Mechanism — 24

    Collapse Of The Wavefunction From The Equation Of Orbital Period Quantizer Mechanism — 27

Evidence Of Orbital Period Quantizer Mechanism Through Planets Of The Solar System — 30

Evidence Of Orbital Period Quantizer Mechanism Through Asteroids Of The Solar System   45

Evidence Of Orbital Period Quantizer Mechanism Through Moons Of Earth's Planetary System   55

Evidence Of Orbital Period Quantizer Mechanism Through Moons Of Mars' Planetary System   59

Evidence Of Orbital Period Quantizer Mechanism Through Moons Of Jupiter's Planetary System   63

Evidence Of Orbital Period Quantizer Mechanism Through Moons Of Saturn's Planetary System   94

Evidence Of Orbital Period Quantizer Mechanism Through Moons Of Uranus' Planetary System   106

Examination Of Stars Of Milky Way Rotating Disk With Potential Gravitational Energy Level Jump In Orbital Period Creation's Quantizer Mechanism: 114

    Verification Cases of Milky Way Stars That May Reveal Orbital Period Creation's Quantizer Mechanism   114

    Revelation Of Deep Quantizer Mechanism Of Orbital Period Creation For Galactic Rotating Disk Hosted Stars   127

Previous Discovery Claim Of Orbital Velocity-Radius Quantizer Mechanism: 132

    Orbital Velocity-Radius Quantizer Mechanism's Nature   132

    Orbital Velocity-Radius Quantizer Mechanism's Components   133

    Orbital Velocity-Radius Quantizer Mechanism's Equations   134

Born Rule Revealed By Orbital Quantizer Mechanism On Orbital Radius Of Captives: 138

    Born Rule In A Nutshell   138

    Captives' Motions Mirroring Wave-Particle Duality   139

Kepler's Second Law And Its Link With Discovered Orbital Period Quantizer Mechanism And Speed Of Gravity: 141

    Kepler's Second Law In A Nutshell   141

    Gravitational Energy Spreading By Captor-Captive Quantum Axle As Quantum Mechanism Behind Kepler's Second Law   141

    Speed Of Gravity Beyond Speed Of Light Revealed By Quantum Mechanism Behind Kepler's Second Law   146

Kepler's Third Law And Its Link With Discovered Orbital Period-Radius-Velocity Quantizer Mechanism: 149

    Kepler's Third Law In A Nutshell   149

    Gravitational Energy Spreading By Captor-Captive Quantum Axle As Quantum Mechanism Behind Kepler's Third Law   150

Orbital Velocity-Radius Quantizer Mechanism Behind Newton's Law Of Orbital Velocity Of Celestial Objects: 152

    Reason-For-Being Of A Hidden Orbital Velocity-Radius Quantizer Mechanism   152

    Internal Core Functions Of Orbital Velocity-Radius Quantizer Mechanism   152

    Function Of Determination Of Orbital Quantizer Via Speed Of Light To Obtain Gravitationally Suitable Radius   153

Of A Captive

    Function Of Determination Of Orbital Quantizer Via   154
    Captor's Gravitational Energy Potential To Obtain
    Gravitationally Suitable Radius Of A Captive

Evidence Of Orbital Velocity-Radius Quantizer Mechanism   156
Through Planets And Asteroids Of The Sun:

    Evidence Of Orbital Velocity-Radius Quantizer   156
    Mechanism Through Planets of The Solar System

    Evidence Of Orbital Velocity-Radius Quantizer   167
    Mechanism Through Asteroids of The Sun

Evidence Of Orbital Velocity-Radius Quantizer Mechanism   172
Through Earth's Moon And Satellite

Evidence Of Orbital Velocity-Radius Quantizer Mechanism   174
Through Mars' Moons

Evidence Of Orbital Velocity-Radius Quantizer Mechanism   176
Through Jupiter's Moons

Evidence Of Orbital Velocity-Radius Quantizer Mechanism   191
Through Saturn's Moons

Evidence Of Orbital Velocity-Radius Quantizer Mechanism   197
Through Uranus' Moons

Evidence Of Orbital Velocity-Radius Quantizer Mechanism   201
Through TOI-178's Exoplanets

Summary Of Orbital Velocity-Radius-Period-Precession   206
Quantizer Mechanism:

    Reason-For-Being Of Orbital Velocity-Radius-Period-   206
    Precession Quantizer Mechanism

Orbital Quantizer's Reason-For-Being	206
Orbital Quantizer's Apparent Link With Lorentz Factor	208
Roles Of Orbital Quantizer For Orbital Radius, Orbital Velocity, Orbital Period, Orbital Precession And Lorentz Factor	208
Reasoning Path Leading To All Equations Of Orbital Velocity-Radius-Period-Precession Quantizer Mechanism:	**210**
Kepler's Third Law Postulate	210
From Newton's Version Of Kepler's Third Law Equation To Speed-Of-Light-Based Orbital Period Quantizer Equation	211
From Speed-Of-Light-Based Orbital Period Quantizer Equation To Orbital Radius Quantizer Equation	213
From Orbital Radius Quantizer Equation To Speed-Of-Light-Based Orbital Period Quantizer Equation	214
Discovery Of Orbital Period Creation's Quantizer Mechanism	215
Discovery Of Orbital Velocity-Radius-Period-Precession Quantizer Mechanism Generating Discrete Not Random Orbital Period Jumps	216
Basic Star System's Orbital Quantizer Constraints	220
Discovery Claim Of Orbital Quantizer Behind Captives' Perihelion Precession Mechanism:	**222**
Simpler Alternative To Gerber-Einstein Equation Of Perihelion Precession Angle	222
Discovery Of Hidden Orbital Quantizer Inside Gerber-Einstein Equation Of Captives' Perihelion Precession	223

Angle

Working Scientific Data Set: 226

    Basic Constants 226

    Studied Solar Planets' Orbital Period Values 227

    Studied Solar Planets' Mean Orbital Velocity Values 227

    Studied Solar Dwarf Planets' Mean Orbital Velocity Values 228

    Studied Planets' Orbital Radius Values 228

    Studied Planets' Core Radius 233

    Studied Planets' Orbital Circumference Values 233

    Studied Planets' Orbital Eccentricity Values 234

    Studied Sun and Planets' Mass Values 235

    Studied Sun and Planet Core Radius Values 236

    Studied Solar Planets Orbital Resonances 236

    Studied Solar Asteroids' Observed Orbital Radius Values 237

    Studied Solar Asteroids' Observed Orbital Period Values 238

    Studied Solar Asteroids' Observed Orbital Circumference Values 238

    Studied Solar Asteroids' Observed Orbital Velocity Values 239

    Studied Solar Asteroids' Observed Orbital Eccentricity Values 239

    Studied Earth's Satellites' Values 240

Studied Mars' Moons Semi-Axis Values	240
Studied Mars' Moons Orbital Period Values	240
Studied Jupiter's Moons Semi-Axis Values	241
Studied Jupiter's Moons Orbital Period Values	242
Studied Jupiter's Moons Orbital Resonances	243
Studied Saturn's Moons Semi-Axis Values	243
Studied Saturn's Moons Orbital Period Values	244
Studied Saturn's Moons Orbital Resonances	244
Studied Uranus' Moons Semi-Axis Values	245
Studied Uranus' Moons Orbital Period Values	245
Final Conclusion	**246**
Edition Changes Notice	**254**

Introduction

The purpose of the present book is to present some of my latest potential discoveries of quantum laws of gravity.

The discoveries at hand are about the

> Wavefunction and quantum superposition of celestial objects that hide behind Kepler's three cosmic laws.

and

> All of these potential discovered quantum behaviors of celestial objects are made possible only through my potential discovered orbital velocity-radius-period-precession quantizer mechanism.
>
>> The relevancy of the said mechanism is that celestial captors appear to use it to generate the orbital velocity, orbital radius, orbital period and orbital precession of each of their celestial captives via a single and unique orbital quantizer value.
>>
>> For short, the said mechanism can be called the "orbital quantizer mechanism".

By the same token,

> The said wavefunction and quantum superposition of celestial objects make Einstein's spacetime continuum paradigm irrelevant, as they appear to have the potential to link already discovered classical laws of gravity to quantum mechanics.

Furthermore, the discoveries at hand help to discover that

> The quantum speeds of gravity appear to exist with two values as the cube and the square of the speed of light (c^3 and c^2).
>
> These quantum speeds of gravity should not be confused with the classical speed of gravity expressed by the speed of gravitational waves, as they do not belong to the same application realms of gravity.

As a matter of fact,

> The first major laws of celestial objects' motions had been discovered by German astronomer and mathematician Johannes Kepler in 1609 and 1619.
>
> Kepler's three laws have revealed the classical mechanics of gravity.

They have set the first major milestone on the very long road to a comprehensive understanding of the laws of gravity. Newton's laws have set another major milestone afterwards then so did Einstein at the beginning of the 20th century.

But among these facts, based on my analysis, it appears that

There exists a continuity between Kepler and Newton, but none between Einstein and his predecessors:

On one hand, Newton discovered his laws and equations that could be directly linked to Kepler's laws. On the other hand, Einstein discovered his laws and equations solely from his unique concept of spacetime, but on the contrary, his laws and equations could not be directly linked to Newton nor Kepler's ones.

At this juncture of physics,

We still do not know how to link Einstein's spacetime to quantum gravity despite more than one century of collective efforts. This was the main reason why I chose not to follow the mainstream physicists' spacetime mantra but to go back to where we were before Einstein, and started from there. This book is one of the results I got so far.

Important Notice:

My analyses, findings and discovery claims, wherever mentioned in this book, must be understood as hypothetical, theoretical and potential at this state of my research work.

All discovery claims need time to be proven or disproved by the international scientific community, and mines are no exception.

What I can say with confidence - at the time of this book's publication - is that my discovery claims have the potential to be validated at some time in the future because they are direct interpretations from discovered equations, and the latter are supported by a host of mathematical evidence upon a large set of relevant astronomical data.

As a result, all items in this book that reflect the status of my research work at this point will have associated words "My" or "Mine", and they should not be confused as items of fact or widely or publicly accepted theories.

Naming Conventions In This Book

In order to avoid repetition of commonly used lengthy words, the latter will be substituted by their shorthand words when needed. These shorthand words are presented here:

Captor:
a celestial object of different type that is orbited by one or many other celestial objects.

> A Captor may actually be a set of celestial objects instead of one, but for the simplification purpose, the single captor case will represent all captor composition cases in this book.

Captive:
a celestial object of different type that orbits another celestial object.

V:
Orbital velocity of a captive as actual observed value.

c:
Speed of light in a vacuum.

Discovery Of Orbital Quantizer Mechanism As Gravitational Link In Orbital Velocity-Radius-Period-Precession Determination For Moving Celestial Objects

In order to understand my main discovery claims presented in this book,

It is necessary to understand first what is my discovered cosmic quantum component of

Orbital Quantizer

This is because the orbital quantizer is one of the fundamental elements that link all my related discovered quantum laws of gravity.

I have previously written two relevant books. The first book presents in details:

Two equations that appear to reveal the orbital quantizer mechanism behind the making of orbital velocities and orbital radii of moving celestial objects.

And the related book's name is:

Orbital Velocity-Radius Quantizer Mechanism Hidden Behind Newton-Einstein Gravity

The second book presents in details:

The third equation that appears to reveal the orbital quantizer mechanism behind the perihelion precession mechanism of orbiting celestial objects.

And the related book's name is:

Einstein's Spacetime Belied By A New Short Perihelion Precession Equation

Then in the present book, I will present my new potential discovery of an

Orbital period quantizer mechanism hidden behind Kepler's second and third laws, and Kepler's first law as a direct consequence therefrom.

And this new potential discovery appears to

Consolidate my previous potential discoveries and to form a unified and extensive cosmic quantum mechanism called here as the orbital velocity-radius-period-precession quantizer mechanism that hides behind Kepler-Newton-Einstein laws of gravity.

By the same token, all these potential discoveries

Consolidate the existence of an even more fundamental quantum cosmic law component under the name of "orbital quantizer".

The "orbital quantizer" mechanism so far appears to be cosmically ubiquitous via two separate sets of gravitational domains:

Orbital velocities, orbital radii, orbital periods and perihelion precessions of moving celestial objects.

Lorentz factor in Einstein's special relativity theory

The core nature of orbital quantizers gives the frequency-like behavior to moving celestial objects.

In order to show the continuity of all my potential discoveries so far, in the present book, the content focus is primarily about

My discoveries of the orbital quantizer mechanism for orbital periods of celestial objects and the frequency nature thereof,

The reason why the orbital quantizer mechanism appears to be somewhat ubiquitous with respect to the apparent harmonious coexistence of captives inside their cosmic systems, and that regardless of the latter's size and structural nature.

The reason why the potential wavefunction, quantum superposition of moving celestial objects and their wavefunction collapse would eventually show up along their orbits and radii.

And finally

The reason why the orbital quantizer paradigm does not need the spacetime continuum paradigm as far as gravity's core mechanism is concerned.

Orbital Quantizer In A Nutshell

Based on my findings,

The equation of the orbital quantizer (denoted originally for short as "Dc" or Divisor of speed of light) is defined as:

$$Dc = c / V \tag{s1}$$

where:
c is the speed of light in vacuum

V is the actual orbital velocity of a specific captive around its captor.

In order to assure the logic consistency throughout my many related books, the "Dc" term - originally standing for "Divisor of speed of light" from my first finding - is maintained instead of other denotations such as "Oq" so one can recognize the same "orbital quantizer" concept in many gravitational equations presented in different books.

There are two automatic derived equations that link "V", "Dc" with the speed-of-light:

$$c = V \times Dc \tag{s4}$$

$$V = c / Dc \tag{s5}$$

The orbital quantizer is a quantum cosmic component of quantization nature.

The orbital quantizer is mathematically a unique multiplier that is attributed by a captor to each of its captives in order to determine the latter's orbital velocity, orbital radius, orbital period and orbital precession.

The orbital quantizer (Dc) operates simultaneously distinctive roles, and these roles are revealed mathematically as follows:

To generate the orbital velocity of a captive,

the orbital quantizer (Dc) becomes the divisor with speed of light as the dividend via the equation (s5):

$$V = c / Dc \tag{=s5}$$

To generate the orbital radius of the same captive,

the orbital quantizer (Dc) becomes the base with exponent of value 2 via the equation (y2):

$$R = [GM/c^2] \times Dc^2 \tag{=y2}$$

The full description of this equation will be presented later on in its dedicated chapter.

To generate the orbital period of the same captive,

the orbital quantizer (Dc) becomes the base with exponent of value 3 via the equation (j2):

$$T = 2\pi \times ([GM/c^3] \times Dc^3) \quad (=j2)$$

The full description of this equation will be presented later on in its dedicated chapter.

To generate the orbital precession (hence precession angle at perihelion) of the same captive,

the orbital quantizer (Dc) becomes the divisor with exponent of value 2 and along with 6π as the dividend via the equation (d1.0):

$$\varepsilon = (6\pi / Dc^2) / (1-e^2) \quad (=d1.0)$$

The full description of this equation will be presented later on in its dedicated chapter.

The orbital quantizer is one of the fundamental reasons why the orbital velocity, orbital radius, orbital period and orbital precession of every captive with respect to the same captor are not random values that gravity somehow sets up by trials and errors. On the contrary, the value of said orbital quantizer is instead a discrete value obtained from the same quantization mechanism/law of gravity applied to a specific celestial object of a specific cosmic system (such as star system, planetary system, moon system and so on).

Once the said orbital quantizer Dc of a captive is known, one can deduce immediately the orbital velocity V, the orbital radius R, the orbital period T and the orbital precession angle ε of this captive.

The one-on-one relationship between the orbital quantizer and the deduced classical orbital velocity "V" of a captive along its orbit around its captor can be shown in the following diagram:

$$V = \sqrt{\frac{GM}{R}} \longrightarrow V = \frac{c}{Dc} \longleftrightarrow Dc = n$$

The said orbital quantizer Dc is hidden in many already discovered physics equations, especially Newton's law of orbital velocity of captives ($V^2=GM/R$), Kepler's second law, Kepler's third law, Einstein's perihelion precession angle equation and all physics equations that include the term "v/c" in them like the Lorentz factor.

Frequency And Wave As Other Fundamental Nature Of Orbital Quantizer

Based on my findings, it appears that

> The equation ($Dc = c / V$) of any orbital quantizer is identical to the equation of any frequency.

Therefore, the two identities " quantizer" and "frequency" can be interchangeable:

> **Orbital Quantizer = Orbital Frequency**

> Every celestial object is quantumly equipped with a specific orbital quantizer, hence with a specific orbital frequency.

> By the same token, every celestial object must vibrate at different specific frequencies at once, in order to maintain its specific orbital radius, orbital velocity, orbital period and orbital precession constant at all times.

Newton's law of orbital velocity of any captive ($V^2=GM/R$) has revealed that the orbital velocity of any captive depends solely on the gravitational mass "GM" value of the captor and the captive's orbital radius.

> Contrary to the value of "GM" that cannot change, as a function of the gravitational mass M of the captor, the orbital radius R of the captive can change, and as a result appears to be a random value. But that is just an illusion. My related discovered equations have contrarily revealed that not only the orbital radius of the captive but the latter's orbital velocity, orbital period and orbital pression are not random values either.

> All following cosmic characteristics of a captive - orbital velocity, orbital radius, orbital period, orbital precession, - are in fact a function of a single cosmic quantizer from gravity laws, hence also a function of orbital frequency or gravitational frequency.

One additional revelation that comes from the quantum object of orbital quantizer is that

> The orbital quantizer appears to be a fundamental link between quantum mechanics and classical mechanics as far as orbital motions are concerned.

> This is because

>> Orbital quantizers and orbital velocities mirror each other while belonging to different mechanics: quantum vs. classical.

>> In quantum mechanics, orbital quantizers are equivalent to orbital

frequencies. They operate in discrete values and behave like waves.

In classical mechanics, orbital velocities operate in continuous values, as objects can accelerate or decelerate, and have no wave behaviors at all.

Discovery Claim Of Orbital Period Quantizer Mechanism Behind Kepler's Cosmic Laws

Kepler's relevant law here is Kepler's third law which states that

> The squares of the orbital periods of the planets are directly proportional to the cubes of the semi-major axes of their orbits.

In this chapter, I will present in details my discovered

> Orbital period quantizer mechanism for celestial objects
>
> > and the equation thereof.
>
> That handles the quantum side of Kepler's third law.

Speed-Of-Light-Based Orbital Period Quantizer Mechanism's Equations

My new claimed discovery pertains to

> A specific orbital period quantizer mechanism that any captor uses to generate a unique speed-of-light-based orbital period for a unique captive.

> This orbital period quantizer mechanism is an integral part of the orbital velocity-radius-period-precession quantizer mechanism for a unique captive.

> The equation of this specific orbital period quantizer mechanism for a specific captive with respect to its captor is defined as follows:

$$T = 2\pi \times ([GM/c^3] \times Dc^3) \qquad (j2)$$

> > where:
> > T is the orbital period of a given captive
> > c is the speed of light in vacuum
> > G is Newton's gravitational constant
> > M is the mass of the captor.
> > Dc is the orbital quantizer created by the said captor for this captive.

This equation (j2) automatically generates two following subsidiary ones:

$$T \times c / 2\pi = [GM/c^2] \times Dc^3 \qquad (j1)$$

$$T / Dc^3 = [GM/c^3] \times 2\pi \qquad (j3)$$

$$T \times c = 2\pi \times ([GM/c^2] \times Dc^3) \qquad (j4)$$

The three equations above (j1, j2, j3 and j4) are intertwined because one equation can be derived from the other, and vice versa. Their differences rely only on the focused purpose or meaning of each equation.

From the (j3), one can deduce the mass of the captor as:

$$M = [T \times c^3] / [Dc^3 \times 2\pi \times G] \qquad (j5)$$

These equations above show how a captor can finalize an orbital quantizer among all available ones that must satisfy the three following equations which were deduced from my discovered orbital velocity-radius-period-precession quantizer mechanism:

$$Dc = [c / V] = \sqrt{(R / [GM/c^2])} \qquad \text{(from (s1))}$$

    As orbital quantizer of power 1 dedicated to orbital velocity of the captive

$$Dc^2 = R / [GM/c^2] \qquad \text{(from (y2))}$$

    As orbital quantizer of power 2 dedicated to orbital radius of the captive

    For orbital radius of captive

$$Dc^3 = T / (2\pi \times [GM/c^3]) \qquad \text{(from (j2))}$$

    As orbital quantizer of power 3 dedicated to orbital period of the captive

We will examine in the next chapter how the fundamental equation (j1) was discovered.

Deep Correlation Between Orbital Period And Orbital Radius In Quantizer-Based Creation Mechanism

Based on my previous and current discovered equations,

    The quantizer equation dedicated to the determination of the orbital radius R of a specific captive is defined as follows:

$$R = [GM/c^2] \times Dc^2 \qquad (=y2)$$

where:
R is the orbital radius of a specific captive.
M is the mass of the captor.
Dc is the orbital quantizer created by the captor for a specific captive.
G is Newton's gravitational constant
c is the speed of light in vacuum

and

The quantizer equation dedicated to the determination of the orbital period T of a specific captive defined as follows:

$$T = 2\pi \times ([GM/c^3] \times Dc^3) \qquad (=j2)$$

Are deeply correlated.

This is because

The manifestation of the said deep correlation between the two quantizer equations above resides in

The pair of gravitational energy "GM/c^2" with orbital quantizer of power 2 "Dc^2"

(that determines the orbital radius of the involved captive)

and

The pair of gravitational energy "GM/c^3" with orbital quantizer of power 3 "Dc^3"

(that determines the orbital period of the involved captive)

Concretely for every captive,

The gravitational energy level must jump from "GM/c^2" to "GM/c^3" and the orbital quantizer level must jump from "Dc^2" to "Dc^3" when the captor's calculation changes from orbital radius to orbital period.

Unless it is a pure coincidence, one has to conclude that

Such deep correlation between the two quantizer equations enforces the coherent quantization nature of gravity with respect to the determination of the orbital radius, orbital velocity and orbital period for a given captive.

The said quantization nature of gravity is expressed as follows:

As far as the values of the orbital radius and orbital period, the captive appears to be allowed to have only discrete values that must also obey the two said equations (y2) and (j2); no random or continuous values are allowed.

As far as the value of the orbital velocity, the captive appears to be assigned with an orbital frequency with the speed-of-light reference, not with a random value.

All elements of the two said equations invoke the speed of light, therefore independent from any so-called time dilation-based consideration.

The parallel increment of the power signs of the speed-of-light elements (from "c^2" to "c^3") and that of the orbital quantizers (from "Dc^2" to "Dc^3") reveal the typical jump of the quantum mechanism to create a pair of orbital radius and orbital period for a given captive.

Emergent Values Of Gravitational Motion Energy Potential Quanta

Based on my findings, there must exist:

Gravitational motion energy potential quanta of level 1 (denoted as E_{g1}) defined via two kinds of units:

The smallest unit thereof as:

$$E_{g1u} = [GM/c^2] \qquad (z1)$$

The generic unit thereof as:

$$E_{g1} = n \times [GM/c^2] \qquad (z2)$$

    where:
    n is a coefficient (multiplier) of integer value starting from 1

This gravitational motion energy potential quanta is one of the fundamental elements in all my discovered equations that can calculate, for each specific captor, the orbital radius, orbital velocity and orbital period for each of its captives.

This gravitational motion energy potential quanta works with another fundamental element in all these discovered equations: the orbital quantizer, attributed uniquely by the captor to each of its captives.

Gravitational motion energy potential quanta of level 2 (denoted as E_{g2}) defined via two kinds of units:

The smallest unit thereof as:

$$E_{g2u} = [GM/c^3] \qquad (z3)$$

The generic unit thereof as:

$$E_{g2} = n \times [GM/c^3] \qquad (z4)$$

where:
n is a coefficient (multiplier) of integer value starting from 1

E_{g2u}'s kind of unit is:

Quantum energy lifespan unit.

Nature Of Quantum observer-measurer-communicator energy lifespan unit:

Quantum observer-measurer-communicator energy lifespan unit is a motion sampling:

during which gravity's energy spreading operates at the speed of the cube value of the speed of light, hence c^3,

and also

after which a photon achieves a distance of 299,792,458 meters, as the measurement of one second according to the SI measurement standard.

Quantum energy lifespan unit is also the lifespan sampling of a theoretical gravity-owned particle that performs the observer-measurer-communicator role between a pair of captor and captive and that can move at the cube value of the speed of light (denoted as "c^3").

In other words,

When a photon finishes to make a distance of 299,792,458 meters, a gravity-owned/dedicated theoretical particle also finishes to make a distance of $(299,792,458)^3$ meters - hence c^3 - if it is still alive during this motion sampling.

Discovery Claim Of Quantum Superposition And Wavefunction Via Orbital Period-Radius-Velocity Quantizer Mechanism Behind Kepler's Cosmic Laws

Quantum Superposition Revelations From The Equation Of Orbital Period Quantizer Mechanism

Based on my findings,

The discovered equations

$$T \times c / 2\pi = [GM/c^2] \times Dc^3 \qquad (=j1)$$

$$T \times c = 2\pi \times ([GM/c^2] \times Dc^3) \qquad (=j4)$$

where:
T is the orbital period of a specific captive.
c is the speed of light in vacuum
G is Newton's gravitational constant
M is the mass of the captor.
Dc is the orbital quantizer created by the captor for a specific captive.

reveals that:

For each captive of the same captor, there must exist at least one quantum twin moving at the speed-of-light along its orbit around the captor, hence the term "T x c".

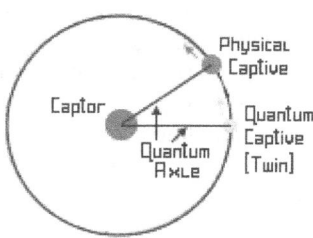

And the said captor uses this quantum twin to attribute a specific orbital period of the said captive using the equations above.

The gravitational energy quanta unit (E_{g1u}) that the captor uses to determine the appropriate orbital period for a specific captive must be "$1GM/c^2$".

The determination of the appropriate orbital period for a specific captive must use the orbital quantizer to the power of 3 hence the term "Dc^3" while that of the orbital radius uses coherently the same orbital quantizer to the power of 2 hence the term "Dc^2":

The term "2π" here is no other than the ratio between two traveling durations/periods at the speed of light, in which

> one test particle goes from end to end along a quantum axle that makes up the orbital radius of the captive,
>
> and the other test particle goes along the captive's orbit to finish one revolution.
>
> In other words, for a specific captive, the term "2π" is the ratio between the period to finish one circular orbit and the period to finish one orbital radius.

Through this orbital period quantizer mechanism:

> The captor can subdivide the speed-of-light by the ratio of 2π to synchronize the quantum orbital radius with the quantum orbit.
>
> Two test particles (one on the quantum orbit and one on the quantum radius) can be triggered at the same time with
>
>> The first test particle located at the captor's center of mass and the second test particle located at the captive's center of mass.
>>
>> These two test particles will end up meeting each other when one finishes its journey along the quantum orbit, and the other finishes its journey along the quantum radius.
>>
>> At this juncture, the captive just finishes one orbital revolution around its captor.
>
> For every specific captive of the same captor, the orbital quantizer is the same quantum element that single-handedly determines both its orbital radius, orbital period and orbital velocity.
>
> The value of the orbital quantizer is unique for a specific captive. This means that no other captives of the same captor can have the same orbital quantizer value unless they share safely the same

orbit.

For the same captive, the cube value of its orbital quantizer (Dc^3) is used to the determine its orbital period while the square value thereof (Dc^2) is used to determine its orbital radius, as shown below:

$$[GM/c^2] \times Dc^3 \qquad \text{(from j1)}$$

vs.

$$[GM/c^2] \times Dc^2 \qquad \text{(from y2)}$$

The combination of the physical captive and its quantum twin appear to reveal the quantum superposition nature of the orbital motion of a captive around its captor.

This is because:

The discovered equation

$$T \times c = 2\pi \times ([GM/c^2] \times Dc^3) \qquad (=j4)$$

hints a permanent presence of a quantum twin moving along the orbit of the captive around its captor at all times.

Let's assume that the captive and its quantum twin each have a quantum axle that connects them to the captor.

By the same token,

> The quantum axle of the said quantum twin keeps rotating constantly around the captor at the speed of light.
>
> The quantum axle of the said physical captive keeps rotating constantly around the captor at its frequency-assigned speed, defined by the assigned orbital quantizer.
>
> When the physical captive achieves one full revolution around the captor, the said quantum twin must finish "Dc" times the same orbital revolution. At that precise moment, the said physical twin must truly collide with the said quantum twin, hence the clash.
>
> And this clash moment can be interpreted as the collapse of the wavefunction of the orbital motion of the physical captive.

Collapse Of The Wavefunction From The Equation Of Orbital Period Quantizer Mechanism

Based on my findings,

The discovered equations

$$T \times c / 2\pi = [GM/c^2] \times Dc^3 \qquad (=j1)$$

$$T \times c = 2\pi \times ([GM/c^2] \times Dc^3) \qquad (=j4)$$

where:
T is the orbital period of a specific captive.
c is the speed of light in vacuum
G is Newton's gravitational constant
M is the mass of the captor.
Dc is the orbital quantizer created by the captor for a specific captive.

reveals that:

The orbital quantizer-generated velocity creates an orbital frequency for each physical captive of the same physical captor.

Therefore, for a unit of orbital motion (i.e. a second in the SI measurements), there are as many quantum axles as the said orbital frequency for a specific physical captive.

During the said unit of orbital motion, whenever a quantum axle collides with the physical captive, it creates a quantum "measurement" and such a "measurement" causes the "collapse of the wavefunction" of the physical captive. And this quantum event causes the orbital location of the physical captive to get redetermined again, hence the latter's advancement along the said orbit.

In this regard,

For a physical captive, its orbital motion is just a continuous series of "collapses of the wavefunction" whose length is determined by the captive's orbital frequency.

As a quantum function, the "collapses of the wavefunction" of physical captives don't depend on their mass scale. This is why stars orbiting black holes can reach velocities near the speed of light.

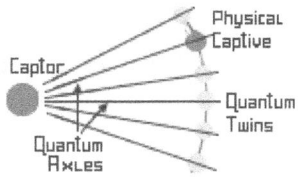

By extrapolation,

> The known phenomena of Einstein rings may be a manifestation of the wavefunction at the galactic scale.

Einstein ring (NASA image)

> Should it be the case, when a gigantic beam of photons approaches a star, the photons thereof would go to the wavefunction rings of this star instead of scattering around. Similar patterns can be seen with the fine ring lines that emerge when a frequency is triggered.

> The known phenomena of asteroid belts (such as Solar Main Asteroid belt, Kuper belt),... may be also manifestations of the wavefunction at the star system scale.

> The known phenomena of planetary asteroid rings (such as Saturn's rings as well as Jupiter, Uranus and Neptune's ones),... may be also manifestations of the wavefunction at the planetary system scale.

Saturn's rings (NASA image)

All these cosmic wavefunction rings are similar to the well known fine structure constant split lines as shown below:

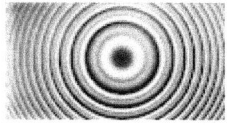

Fire structure constant Split lines (Wikipedia)

Evidence Of Orbital Period Quantizer Mechanism Through Planets Of The Solar System

Targets Of This Orbital Radius Period Quantizer Mechanism Verification:

Based on my calculations, there appears to exist

A host of evidence of this orbital period quantizer mechanism for a host of known planets that revolve around the Sun.

The planets subjected to this verification are:

Mercury, Venus, Earth. Mars, Jupiter, Saturn, Uranus, Neptune, Pluto

For each planet therein, we will verify whether or not

The orbital period of each planet of the Solar system obeys the equation:

$$T = 2\pi \times ([GM/c^3] \times Dc^3) \qquad (=j2)$$

The equation of orbital period (j2) shows that time is no other than a mathematical representation of chain of motions generated by gravitational energy, not a pure dimension as Einstein's spacetime paradigm claims.

Reference Data Of This Verification:

From the equations (z1), (z2) and the values of:

M_{sun} = 1.9885e+30 kg (as mass of the Sun)
G = 6.6743e-11 (N kg^{-2} m^2)
c = 299,792,458 m

we obtain:

The Sun's effective gravitational energy potential quanta dedicated exclusively to orbital motion of each captive as:

GM_{sun} = 1.327184555e+20 N kg^3 m^2
(= 6.6743e-11 kg^{-2}m^2 × 1.9885e+30 kg)

$$GM_{sun}/c^2 = E_{g1} = 1 \times [GM_{sun}/c^2] \qquad (=z1)$$

$GM_{sun}/c^2 = 1,476.6919$ m or 1.476619 km

$GM_{sun}/c^3 = E_{g2} = 1 \times [GM_{sun}/c^3]$ (=z3)

$GM_{sun}/c^3 = 0.000004925714206973968$ seconds

Let's go through each planet subjected to this verification.

Orbital Period Verification Case of Mercury:

From the astronomically

Observed mean orbital velocity of Mercury ($V_{mercury}$) as:

$V_{mercury} = 47.870$ km/s or $47,870$ m/s
Source:

and the equation (s1) we obtain:

Mercury's orbital quantizer value $Dc\_mercury$ as:

$Dc\_mercury = c / V_{mercury} = 299,792,458$ m $/ 47,870$ m $\rightarrow$

$Dc\_mercury = 6,262.62$ $\rightarrow$

$Dc\_mercury^2 = 39,220,409.26$

$Dc\_mercury^3 = 245,622,519,467.41$

That leads to obtain the following Mercury terms of the equation (j2):

$[GM_{sun}/c^3] \times Dc\_mercury^3$ $= 0.000004925714206973968 \times 245,622,519,467.41$
                                            $= 1,209,866.33$ secs

$[GM_{sun}/c^3] \times Dc\_mercury^3 \times 2\pi$ $= 1,209,866.33 \times 2\pi$
                                              $= 7,601,814.34$ seconds

The outcome of the equation (j2) of the orbital period quantizer mechanism for Mercury yields:

Mercury's Predicted orbital period ($Tp\_mercury$)   = **7,601,814.34 seconds**

On the other hand,

Observed orbital period of Mercury ($To\_mercury$) gives:

$To\_mercury$ $= 7,600,530.24$ seconds
                              (or 87.9691 days)

Let's calculate

The percentage margin of error between Mercury's Predicted orbital period and its astronomically observed one:

$\Delta[Tp\_mercury / To\_mercury]$ = [7,601,814.34 secs / 7,600,530.24 secs] x 100
= 0.01689%

One can conclude that

The equation (j2) of the orbital period quantizer mechanism predicts correctly Mercury's orbital period

Orbital Period Verification Case of Venus:

From the astronomically

Observed mean orbital velocity of Venus (Vvenus) as:

Vvenus = 35.020 km/s or 35,020 m/s
Source:

and the equation (s1) we obtain:

Venus' orbital quantizer value Dc_venus as:

Dc_venus = c / Vvenus = 299,792,458 m / 35,020 m →

Dc_venus = 8,560.60 →

Dc_venus^2 = 73,283,872.36 →

Dc_venus^3 = 627,353,917,725.01 →

That leads to obtain the following Venus terms of the equation (j2):

$[GMsun/c^3]$ x Dc_venus^3 = 0.000004925714206973968 x 627,353,917,725.01
= 3,090,166.10 seconds

$[GMsun/c^3]$ x Dc_venus^3 x 2π = 3,090,166.10 x 2π
= 19,416,086.23 seconds

The outcome of the equation (j2) of the orbital period quantizer mechanism for Venus yields:

Venus'
Predicted orbital period
(Tp_venus) = **19,416,086.23 seconds**

On the other hand,

Observed orbital period of Venus (To_venus) gives:

To_venus = 19,414,080 seconds
(or 224.7 days)

Let's calculate

The percentage margin of error between Venus' Predicted orbital period and its astronomically observed one:

$\Delta[Tp\_venus / To\_venus]$ = [19,416,086.23 secs / 19,414,080 secs] x 100
= 0.01033%

One can conclude that

The equation (j2) of the orbital period quantizer mechanism predicts correctly Venus' orbital period

Orbital Period Verification Case of Earth:

From the astronomically

Observed mean orbital velocity of Earth ($Vearth$) as:

$Vearth$ = 29.780 km/s or 29,780 m/s
Source:

and the equation (s1) we obtain:

Earth's orbital quantizer value Dc_earth as:

$Dc\_earth = c / Vearth$ = 299,792,458 m / 29,780 m $\rightarrow$

Dc_earth = 10,066.90 $\rightarrow$

Dc_earth^2 = 101,342,475.61 $\rightarrow$

Dc_earth^3 = 1,020,204,567,718.30 $\rightarrow$

That leads to obtain the following Earth terms of the equation (j2):

$[GMsun/c^3] \times Dc\_earth^3$ = 0.000004925714206973968 x 1,020,204,567,718.30
= 5,025,236.13 seconds

$[GMsun/c^3] \times Dcearth^3 \times 2\pi$ = 5,025,236.13 x 2π
= 31,574,489.81 seconds

The outcome of the equation (j2) of the orbital period quantizer mechanism for Earth yields:

Earth's Predicted orbital period (Tp_earth) = **31,574,489.81 seconds**

On the other hand,

Observed orbital period of Earth (To_earth) gives:

To_earth = 31,558,149.504 seconds (or 365.25636 days)

Let's calculate

The percentage margin of error between Earth's Predicted orbital period and its astronomically observed one:

$\Delta[Tp\_earth / To\_earth]$ = [31,574,489.81 secs / 31,558,149.504 secs] x 100 = 0.05177%

One can conclude that

The equation (j2) of the orbital period quantizer mechanism predicts correctly Earth's orbital period

Orbital Period Verification Case of Mars:

From the astronomically

Observed mean orbital velocity of Mars ($Vmars$) as:

$Vmars$ = 12.4514 km/s or 12,451.40 m/s
Source:

and the equation (s1) we obtain:

Mars' orbital quantizer value Dc_mars as:

$Dc\_mars = c / Vmars$ = 299,792,458 m / 12,451.40 m $\rightarrow$

Dc_mars = 12,451.40 $\rightarrow$

Dc_mars^2 = 155,037,361.96 $\rightarrow$

Dc_mars^3 = 1,930,432,208,708.744 $\rightarrow$

That leads to obtain the following Mars terms of the equation (j2):

$[GM_{sun}/c^3] \times Dc\_mars^3$ $= 0.000004925714206973968 \times 1,930,432,208,708.744$
$= 9,508,757.35$ seconds

$[GM_{sun}/c^3] \times Dc\_mars^3 \times 2\pi$ $= 9,508,757.35 \times 2\pi$
$= 59,745,284.47$ seconds

The outcome of the equation (j2) of the orbital period quantizer mechanism for Mars yields:

Mars' Predicted orbital period (Tp_mars) $= \mathbf{59,745,284.47}$ **seconds**

On the other hand,

Observed orbital period of Mars (To_mars) gives:

To_mars $= 59,360,000$ seconds (or 687.0 days)

Let's calculate

The percentage margin of error between Mars' Predicted orbital period and its astronomically observed one:

$\Delta[Tp\_mars / To\_mars]$ $= 59,745,284.47$ secs / $59,360,000$ secs
$= 0.649\%$

One can conclude that

The equation (j2) of the orbital period quantizer mechanism predicts correctly Mars' orbital period

Orbital Period Verification Case of Jupiter:

From the astronomically

Observed mean orbital velocity of Jupiter ($Vjupiter$) as:

$Vjupiter = 13.070$ km/s or $13,070$ m/s
Source:

and the equation (s1) we obtain:

Jupiter's orbital quantizer value $Dc\_jupiter$ as:

$Dc\_jupiter = c / Vjupiter = 299,792,458$ m / $13,070$ m $\rightarrow$

$Dc\_jupiter = 22,937.44$ $\rightarrow$

$Dc\_jupiter^2 = 526{,}126{,}153.75$ →

$Dc\_jupiter^3 = 12{,}067{,}987{,}084{,}153.97$ →

That leads to obtain the following Jupiter terms of the equation (j2):

$[GM_{sun}/c^3] \times Dc\_jupiter^3$ = $0.0000049257142069739687 \times 12{,}067{,}987{,}084{,}153.97$
= $59{,}443{,}455.43$ seconds

$[GM_{sun}/c^3] \times Dc\_jupiter^3 \times 2\pi$ = $59{,}443{,}455.43 \times 2\pi$
= $373{,}494{,}245.76$ seconds

The outcome of the equation (j2) of the orbital period quantizer mechanism for Jupiter yields:

Jupiter's Predicted orbital period (Tp_jupiter) = **373,494,245.76 seconds**

On the other hand,

Observed orbital period of Jupiter (To_jupiter) gives:

To_jupiter = $374{,}198{,}000$ seconds (or 4,331 days)

Let's calculate

The percentage margin of error between Jupiter's Predicted orbital period and its astronomically observed one:

$\Delta[Tp\_jupiter / To\_jupiter]$ = $[373{,}494{,}245.76 \text{ secs} / 374{,}198{,}000 \text{ secs}] \times 100$
= 0.1880%

One can conclude that

The equation (j2) of the orbital period quantizer mechanism predicts correctly Jupiter's orbital period

Orbital Period Verification Case of Saturn:

From the astronomically

Observed mean orbital velocity of Saturn (Vsaturn) as:

Vsaturn = 9.690 km/s or 9,690 m/s
Source:

and the equation (s1) we obtain:

Saturn's orbital quantizer value Dc_saturn as:

$Dc\_saturn = c / Vsaturn = 299,792,458 \text{ m} / 9,690 \text{ m}$ →

$Dc\_saturn = 30,938.33$ →

$Dc\_saturn^2 = 957,180,263.18$ →

$Dc\_saturn^3 = 29,613,558,852,025.04$ →

That leads to obtain the following Saturn terms of the equation (j2):

$[GMsun/c^3] \times Dc\_saturn^3$	$= 0.000004925714206973968 \times 29,613,558,852,025.04$ $= 145,867,927.55$ seconds
$[GMsun/c^3] \times Dc\_saturn^3 \times 2\pi$	$= 145,867,927.55 \times 2\pi$ $= 916,515,219.17$ seconds

The outcome of the equation (j2) of the orbital period quantizer mechanism for Saturn yields:

Saturn's Predicted orbital period (Tp_saturn)　　**= 916,515,219.17 seconds**

On the other hand,

Observed orbital period of Saturn (To_saturn) gives:

To_saturn　　$= 928,540,800$ seconds (or 10,747 days)

Let's calculate

The percentage margin of error between Saturn's Predicted orbital period and its astronomically observed one:

$\Delta[Tp\_saturn / To\_saturn]$　　$= [916,515,219.17 \text{ secs} / 928,540,800 \text{ secs}] \times 100$
$= 1.2951\%$

One can conclude that

The equation (j2) of the orbital period quantizer mechanism predicts correctly Saturn's orbital period

Orbital Period Verification Case of Uranus:

From the astronomically

Observed mean orbital velocity of Uranus ($Vuranus$) as:

Vuranus = 6.810 km/s or 6,810 m/s
Source:

and the equation (s1) we obtain:

Uranus' orbital quantizer value Dc_uranus as:

$Dc\_uranus = c / Vuranus = 299{,}792{,}458$ m $/ 6{,}810$ m   →

with $Vuranus$ as Uranus' observed orbital velocity   →

$Dc\_uranus = 44{,}022.38$ →

$Dc\_uranus^2 = 1{,}937{,}969{,}940.86$ →

$Dc\_uranus^3 = 85{,}314{,}049{,}165{,}310.14$ →

That leads to obtain the following Uranus terms of the equation (j2):

$[GMsun/c^3] \times Dc\_uranus^3$	$= 0.000004925714206973968 \times 85{,}314{,}049{,}165{,}310.14$ $= 420{,}232{,}624.02$ seconds
$[GMsun/c^3] \times Dc\_uranus^3 \times 2\pi$	$= 420{,}232{,}624.02 \times 2\pi$ $= 2{,}640{,}399{,}448.84$ seconds

The outcome of the equation (j2) of the orbital period quantizer mechanism for Uranus yields:

Uranus' Predicted orbital period (Tp_uranus)   **= 2,640,399,448.84 seconds**

On the other hand,

Observed orbital period of Uranus (To_uranus) gives:

To_uranus $= 2{,}642{,}890{,}000$ seconds
(or 30,589 days)

Let's calculate

The percentage margin of error between Uranus' Predicted orbital period and its astronomically observed one:

$\Delta[Tp\_uranus / To\_uranus]$ $= [2{,}640{,}399{,}448.84$ secs $/ 2{,}642{,}890{,}000$ secs$] \times 100$
$= 0.0942\%$

One can conclude that

The equation (j2) of the orbital period quantizer mechanism predicts correctly Uranus' orbital period

Orbital Period Verification Case of Neptune:

From the astronomically

Observed mean orbital velocity of Neptune (Vneptune) as:

Vneptune = 5.430 km/s or 5,430 m/s
Source:

and the equation (s1) we obtain:

Neptune's orbital quantizer value Dc_neptune as:

Dc_neptune = c / Vneptune = 299,792,458 m / 5,430 m $\rightarrow$

Dc_neptune = 55,210.39 $\rightarrow$

Dc_neptune^2 = 3,048,187,163.95 $\rightarrow$

Dc_neptune^3 = 168,291,602,114,789.38 $\rightarrow$

That leads to obtain the following Neptune terms of the equation (j2):

[GMsun/c^3] x Dc_neptune^3 = 0.000004925714206973968 x 168,291,602,114,789.38
= 828,956,335.45 seconds

[GMsun/c^3] x Dc_neptune^3 x 2π = 828,956,335.45 x 2π
= 5,208,486,267.19 seconds

The outcome of the equation (j2) of the orbital period quantizer mechanism for Neptune yields:

Neptune's Predicted orbital period (Tp_neptune) = **5,208,486,267.19 seconds**

On the other hand,

Observed orbital period of Neptune (To_saturn) gives:

To_neptune = 5,166,720,000 seconds
(or 59,800 days)

Let's calculate

The percentage margin of error between Neptune's Predicted orbital period and its astronomically observed one:

Δ[Tp_neptune / To_neptune] = [5,208,486,267.19 secs / 5,166,720,000 secs] x 100

= 0.8083%

One can conclude that

The equation (j2) of the orbital period quantizer mechanism predicts correctly Neptune's orbital period

Orbital Period Verification Case of Pluto:

From the astronomically

Observed mean orbital velocity of Pluto (V_{pluto}) as:

V_{pluto} = 4.740 km/s or 4,740 m/s
Source:

and the equation (s1) we obtain:

Pluto's orbital quantizer value Dc_pluto as:

$Dc\_pluto = c / V_{pluto}$ = 299,792,458 m / 4,740 m →

Dc_pluto = 63,247.35 →

Dc_pluto^2 = 3,048,187,163.95 →

Dc_pluto^3 = 253,003,774,985,625.76 →

That leads to obtain the following Pluto terms of the equation (j2):

$[GM_{sun}/c^3] \times Dc\_pluto^3$	= 0.000004925714206973968 x 253,003,774,985,625.76 = 1,246,224,288.86 seconds
$[GM_{sun}/c^3] \times Dc\_pluto^3 \times 2\pi$	= 1,246,224,288.86 x 2π = 7,830,258,141.21 seconds

The outcome of the equation (j2) of the orbital period quantizer mechanism for Pluto yields:

Pluto's Predicted orbital period (Tp_pluto)   = **7,830,258,141.21 seconds**

On the other hand,

Observed orbital period of Pluto (To_pluto) gives:

To_pluto = 7,824,380,000 seconds
(or 90,560 days)

Let's calculate

The percentage margin of error between Pluto's Predicted orbital period and its astronomically observed one:

$\Delta[Tp\_pluto / To\_pluto]$ = [7,830,258,141.21 secs / 7,824,380,000 secs] x 100
= 0.0751%

One can conclude that

The equation (j2) of the orbital period quantizer mechanism predicts correctly Pluto's orbital period

Orbital Period Verification Case of Eris:

From the astronomically

Observed mean orbital velocity of Eris (Veris) as:

Veris = 3.434 km/s or 3,434 m/s
Source:

and the equation (s1) we obtain:

Eris' orbital quantizer value Dc_eris was known as:

Dc_eris = c / Veris = 299,792,458 m / 3,434 m   →

Dc_eris = 87,301.24   →

Dc_eris^2 = 7621506505.53   →

Dc_eris^3 = 665,366,968,601,499.34   →

That leads to obtain the following Eris terms of the equation (j2):

[GMsun/c^3] x Dc_eris^3 = 0.000004925714206973968 x 665,366,968,601,499.34
= 3,277,407,530.09 seconds

[GMsun/c^3] x Dc_eris^3 x 2π = 3,277,407,530.09 x 2π
= 20,592,558,838.70 seconds

The outcome of the equation (j2) of the orbital period quantizer mechanism for Eris yields:

Eris' Predicted orbital period (Tp_eris)   = **20,592,558,838.70 seconds**

On the other hand,

Observed orbital period of Eris (To_eris) gives:

T_{o_eris} = 17,630,000,000 seconds
(or 559 years)

Let's calculate

The percentage margin of error between Eris' Predicted orbital period and its astronomically observed one:

T_{p_eris} / T_{o_eris} = [20,592,558,838.70 secs / 17,630,000,000 secs] x 100
= 16.80%

One can conclude that

The equation (j2) of the orbital period quantizer mechanism predicts correctly Eris' orbital period

Orbital Period Verification Case of Sedna:

From the astronomically

Observed mean orbital velocity of Sedna (V_{sedna}) as:

V_{sedna} = 1.04 km/s or 1,040 m/s
Source:

and the equation (s1) we obtain:

Sedna' orbital quantizer value D_{c_sedna} was known as:

D_{c_sedna} = c / V_{sedna} = 299,792,458 m / 1,040 m →

D_{c_sedna} = 288,261.97 →

$D_{c\_sedna}^2$ = 83,094,963,348.2809 →

$D_{c\_sedna}^3$ = 23,953,117,831,853,248 →

That leads to obtain the following Sedna terms of the equation (j2):

$[GM_{sun}/c^3]$ x $D_{c\_sedna}^3$ = 0.000004925714206973968 x 23,953,117,831,853,248
= 117,986,212,805.68 seconds

$[GM_{sun}/c^3]$ x $D_{c\_sedna}^3$ x 2π = 117,986,212,805.68 x 2π
= 741,329,238,750.41 secs

The outcome of the equation (j2) of the orbital period quantizer mechanism for Sedna yields:

Sedna' Predicted orbital period   = **741,329,238,750.41 secs**
(T_{p_sedna})   **(or 23,507.39 years)**

On the other hand,

Observed orbital period of Sedna (To_sedna) gives:

$$\text{To\_sedna} = 359{,}510{,}000{,}000 \text{ seconds}$$
(or 11,400 years) or

$$= 354{,}748{,}500{,}000 \text{ seconds}$$
(or 11,249 years) or

Let's calculate

The percentage margin of error between Sedna' Predicted orbital period and its astronomically observed one:

$$\text{Tp\_sedna / To\_sedna} = [741{,}329{,}238{,}750.41 \text{ secs} / 354{,}748{,}500{,}000 \text{ secs}] \times 100$$
$$= 208.9731\% \; ????$$

One can conclude that

The equation (j2) of the orbital period quantizer mechanism predicts correctly Sedna' orbital period

Orbital Period Verification Case of Goblin (The):

As of 2023, there is no available data about the observed orbital velocity of the Goblin. The only available data (from Theskylive.com) are:

The Goblin's Orbital Radius (From ~ 1,369.79770882 AU)	204,918,820,941 km
The Goblin's Perihelion Radius (From ~ 65.11260610 AU)	9,740,707,247 km
The Goblin's Aphelion Radius (From ~ 2,674.48281155 AU)	400,096,934,634 km
The Goblin's Orbital Period	40,000 Earth-years or 1.26144e+12 seconds

From the astronomically

Observed mean orbital velocity of Goblin (The) (Vgoblin) as:

Vgoblin = 1.02 km/s or 1,020 m/s
(From 204,918,820,941 km x 2π / 1.26144e+12 seconds)

and the equation (s1) we obtain:

Goblin (The)' orbital quantizer value Dc_goblin was known as:

$Dc\_goblin = c / V_{goblin} = 299{,}792{,}458 \text{ m} / 1{,}020 \text{ m}$ →

$Dc\_goblin = 293{,}914.17$ →

$Dc\_goblin^2 = 86{,}385{,}539{,}326.7889$ →

$Dc\_goblin^3 = 25{,}389{,}934{,}091{,}235{,}518$ →

That leads to obtain the following Goblin (The) terms of the equation (j2):

$[GM_{sun}/c^3] \times Dc\_goblin^3$ = 0.000004925714206973968 x 25,389,934,091,235,518
= 125,063,559,067.33 seconds

$[GM_{sun}/c^3] \times Dc\_goblin^3 \times 2\pi$ = 125,063,559,067.33 x 2π
= 785,797,516,795.43 secs

The outcome of the equation (j2) of the orbital period quantizer mechanism for Goblin (The) yields:

Goblin (The)' Predicted orbital period (Tp_goblin) = **785,797,516,795.43 secs (or 24,917.47 years)**

On the other hand,

Observed orbital period of Goblin (The) (To_goblin) gives:

To_goblin = 1,261,440,000,000 seconds (or 40,000 years)

Let's calculate

The percentage margin of error between Goblin (The)' Predicted orbital period and its astronomically observed one:

Tp_goblin / To_goblin = [785,797,516,795.43 secs / 1,261,440,000,000.00 secs] x 100
= 160.53% ????

One can conclude that

The equation (j2) of the orbital period quantizer mechanism predicts correctly Goblin (The)' orbital period

Evidence Of Orbital Period Quantizer Mechanism Through Asteroids Of The Solar System

Targets Of This Orbital Radius Period Quantizer Mechanism Verification:

Based on my calculations, there appears to exist

> A host of evidence of this orbital period quantizer mechanism for a host of known asteroids that revolve around the Sun.

The asteroids subjected to this verification are:

> Ra-shalom, Aten, Icarus, Phaethon, Toro, Apollo, Adonis

For each asteroid therein, we will verify whether or not

> The orbital period of each well-known asteroid of the Solar system obeys the equation:
>
> $$T = 2\pi \times ([GM/c^3] \times Dc^3) \qquad (=j2)$$

Reference Data Of This Orbital Period Verification:

From the equations (z1), (z2) and the values of:

> M_{sun} = 1.9885e+30 kg (as mass of the Sun)
> G = 6.6743e-11 (N kg^{-2} m²)
> c = 299,792,458 m

we obtain:

> The Sun's effective gravitational energy potential quanta dedicated exclusively to orbital motion of each captive as:
>
> GM_{sun} = 1.327184555e+20 N kg³ m²
>   (= 6.6743e-11 kg^{-2}m² x 1.9885e+30 kg)
>
> $GM_{sun}/c^2 = E_{g1} = 1 \times [GM_{sun}/c^2]$ $\qquad (=z1)$
>
> GM_{sun}/c^2 = 1,476.6919 m or 1.476619 km
>
> $GM_{sun}/c^3 = E_{g2} = 1 \times [GM_{sun}/c^3]$ $\qquad (=z3)$

$GM_{sun}/c^3 = 0.000004925714206973968$ seconds

Let's go through each asteroid subjected to this verification.

Orbital Period Verification Case of Asteroid of Ra-shalom:

From the astronomically

Observed mean orbital velocity of Ra-shalom ($V_{rashalom}$) as:

$V_{rashalom}$ = 32.652 km/s or 32,652 m/s
Source:

and the equation (s1) we obtain:

Ra-shalom's orbital quantizer value $Dc\_rashalom$ was known as:

$Dc\_rashalom = c / V_{rashalom}$ = 299,792,458 m / 32,652 m →

$Dc\_rashalom$ = 9,181.44 →

$Dc\_rashalom^2$ = 84,298,840.47 →

$Dc\_rashalom^3$ = 773,984,745,877.92 →

That leads to obtain the following Ra-shalom terms of the equation (j2):

$[GM_{sun}/c^3]$ x $Dc\_rashalom^3$ = 0.000004925714206973968 x 773,984,745,877.92
= 3,812,427.65 seconds

$[GM_{sun}/c^3]$ x $Dc\_rashalom^3$ x 2π = 3,812,427.65 x 2π
= 23,954,189.39 seconds

The outcome of the equation (j2) of the orbital period quantizer mechanism for Ra-shalom yields:

Ra-shalom' Predicted orbital period ($Tp\_rashalom$) = **23,954,189.39 seconds**

On the other hand,

Observed orbital period of Ra-shalom ($To\_rashalom$) gives:

$To\_rashalom$ = 23,951,182 seconds
(or 0.759487 years)

Source:

Let's calculate

The percentage margin of error between Ra-shalom's Predicted orbital period and its astronomically observed one:

$\Delta[\text{Tp\_rashalom} / \text{To\_rashalom}]$ = [23,954,189.39 secs / 23,951,182 secs] x 100
= 0.0125%

One can conclude that

The equation (j2) of the orbital period quantizer mechanism predicts correctly Ra-shalom's orbital period

Orbital Period Verification Case of Asteroid of Aten:

From the astronomically

Observed mean orbital velocity of Aten (Vaten) as:

Vaten = 30.336 km/s or 30,336 m/s
Source:

and the equation (s1) we obtain:

Aten's orbital quantizer value Dc_aten as:

$Dc\_aten = c / Vaten = 299{,}792{,}458 \text{ m} / 30{,}336 \text{ m}$ →

$Dc\_aten = 9{,}882.39$ →

$Dc\_aten^2 = 97{,}661{,}632.11$ →

$Dc\_aten^3 = 965{,}130{,}336{,}568.29$ →

That leads to obtain the following Aten terms of the equation (j2):

$[GM_{sun}/c^3] \times Dc\_aten^3$ = 0.000004925714206973968 x 965,130,336,568.29
= 4,753,956.21 seconds

$[GM_{sun}/c^3] \times Dc\_aten^3 \times 2\pi$ = 4,753,956.21 x 2π
= 29,869,987.80 seconds

The outcome of the equation (j2) of the orbital period quantizer mechanism for Aten yields:

Aten's Predicted orbital period (Tp_aten) = **29,869,987.80 seconds**

On the other hand,

Observed orbital period of Aten (T_{o_aten}) gives:

T_{o_aten} = 29,959,000 seconds
(or 0.950 years)

Source:
https://academickids.com/encyclopedia/index.php/1566_Aten#google_vignette

Let's calculate

The percentage margin of error between Aten's predicted orbital period and its astronomically observed one:

$\Delta[T_{p\_aten} / T_{o\_aten}]$ = [29,869,987.80 secs / 29,959,000 secs] x 100
= 0.297%

One can conclude that

The equation (j2) of the orbital period quantizer mechanism predicts correctly Aten's orbital period

Orbital Period Verification Case of Asteroid of Icarus:

From the astronomically

Observed mean orbital velocity of Icarus (V_{icarus}) as:

V_{icarus} = 28.69 km/s or 28,690 m/s
Source:
https://www.spacereference.org/asteroid/1566-icarus-1949-ma
https://academickids.com/encyclopedia/index.php/1566_Icarus#google_vignette

and the equation (s1) we obtain:

Icarus' orbital quantizer value D_{c_icarus} as:

D_{c_icarus} = c / V_{icarus} = 299,792,458 m / 28,690 m   →

D_{c_icarus} = 10,449.37   →

$D_{c\_icarus}^2$ = 109,189,333.39   →

$D_{c\_icarus}^3$ = 1,140,959,744,717.56   →

That leads to obtain the following Icarus terms of the equation (j2):

$[GM_{sun}/c^3]$ x $D_{c\_icarus}^3$ = 0.000004925714206973968 x 1,140,959,744,717.56
= 5,620,041.62 seconds

$[GM_{sun}/c^3] \times Dc\_icarus^3 \times 2\pi$ $= 5{,}620{,}041.62 \times 2\pi$
$= 35{,}311{,}762.93$ seconds

The outcome of the equation (j2) of the orbital period quantizer mechanism for Icarus yields:

Icarus' Predicted orbital period (Tp_icarus) = **35,311,762.93 seconds**

On the other hand

Observed orbital period of Icarus (To_icarus) gives:

To_icarus = 35,318,419.2 seconds
(or 408.778 days
or 1.119939 years)

Source: https://academickids.com/encyclopedia/index.php/1566_Icarus#google_vignette

Let's calculate

The percentage margin of error between Icarus' predicted orbital period and its astronomically observed one:

$\Delta[Tp\_icarus / To\_icarus]$ = [35,311,762.93 secs / 35,318,419.2 secs] x 100
= 0.0188464%

One can conclude that

The equation (j2) of the orbital period quantizer mechanism predicts correctly Icarus' orbital period

Orbital Period Verification Case of Asteroid of Phaethon:

From the astronomically

Observed mean orbital velocity of Phaethon (Vphaethon) as:

Vphaethon = 26.415 km/s or 26,415 m/s
Source:

and the equation (s1) we obtain:

Phaethon's orbital quantizer value Dc_phaethon as:

Dc_phaethon = c / Vphaethon = 299,792,458 m / 26,415 m →

Dc_phaethon = 11,349.32 →

$Dc\_phaethon^2 = 128{,}807{,}064.46$ →

$Dc\_phaethon^3 = 1{,}461{,}872{,}592{,}844.40$ →

That leads to obtain the following Phaethon terms of the equation (j2):

$[GMsun/c^3] \times Dc\_phaethon^3$ = 0.000004925714206973968 x 1,461,872,592,844.40
= 7,200,766.59 seconds

$[GMsun/c^3] \times Dc\_phaethon^3 \times 2\pi$ = 7,200,766.59 x 2π
= 45,243,750.83 seconds

The outcome of the equation (j2) of the orbital period quantizer mechanism for Phaethon yields:

Phaethon's Predicted orbital period ($Tp\_phaethon$) = **45,243,750.83 seconds**

On the other hand,

Observed orbital period of Phaethon ($To\_phaethon$) gives:

$To\_phaethon$ = 45,241,545 seconds
(or 1.4346 years)

Source:

Let's calculate

The percentage margin of error between Phaethon's Predicted orbital period and its astronomically observed one:

$\Delta[Tp\_phaethon / To\_phaethon]$ = [45,243,750.83 secs / 45,241,545 secs] x 100
= 0.00487%

One can conclude that

The equation (j2) of the orbital period quantizer mechanism predicts correctly Phaethon's orbital period

Orbital Period Verification Case of Asteroid of Toro:

From the astronomically

Observed mean orbital velocity of Toro ($Vtoro$) as:

$Vtoro$ = 25.471 km/s or 25,471 m/s
Source:

and the equation (s1) we obtain:

Toro's orbital quantizer value Dc_toro was known as:

$Dc\_toro = c / Vtoro = 299{,}792{,}458 \text{ m} / 25{,}471 \text{ m}$ →

$Dc\_toro = 11{,}769.95$ →

$Dc\_toro^2 = 138{,}531{,}723.00$ →

$Dc\_toro^3 = 1{,}630{,}511{,}453{,}153.27$ →

That leads to obtain the following Toro terms of the equation (j2):

$[GMsun/c^3] \times Dc\_toro^3$ = 0.0000049257142069973968 x 1,630,511,453,153.27
= 8,031,433.42 seconds

$[GMsun/c^3] \times Dc\_toro^3 \times 2\pi$ = 8,031,433.42 x 2π
= 50,462,984.46 seconds

The outcome of the equation (j2) of the orbital period quantizer mechanism for Toro yields:

Toro's Predicted orbital period (Tp_toro) = **50,462,984.46 seconds**

On the other hand,

Observed orbital period of Toro (To_toro) gives:

To_toro = 50,464,598 seconds
(or 1.599 years)

Source:

Let's calculate

The percentage margin of error between Toro's Predicted orbital period and its astronomically observed one:

$\Delta[Tp\_toro / To\_toro]$ = [50,462,984.46 secs / 50,464,598 secs] x 100
= 0.00319%

One can conclude that

The equation (j2) of the orbital period quantizer mechanism predicts correctly Toro's orbital period

Orbital Period Verification Case of Asteroid of Apollo:

From the astronomically

Observed mean orbital velocity of Apollo (V_{apollo}) as:

V_{apollo} = 24.552 km/s or 24,552 m/s
Source:

and the equation (s1) we obtain:

Apollo's orbital quantizer value Dc_apollo as:

$Dc\_apollo = c / V_{apollo}$ = 299,792,458 m / 24,552 m →

Dc_apollo = 12,210.51 →

Dc_apollo^2 = 149,096,554.46 →

Dc_apollo^3 = 1,820,544,969,200.59 →

That leads to obtain the following Apollo terms of the equation (j2):

$[GM_{sun}/c^3] \times Dc\_apollo^3$ = 0.000004925714206973968 × 1,820,544,969,200.59
= 8,967,484.22 seconds

$[GM_{sun}/c^3] \times Dc\_apollo^3 \times 2\pi$ = 8,967,484.22 × 2π
= 56,344,365.09 seconds

The outcome of the equation (j2) of the orbital period quantizer mechanism for Apollo yields:

Apollo' Predicted orbital period (Tp_apollo) = **56,344,365.09 seconds**

On the other hand,

Observed orbital period of Apollo (To_apollo) gives:

To_apollo = 56,291,760 seconds
(or 1.785 years)

Source:

Let's calculate

The percentage margin of error between Apollo's Predicted orbital period and its astronomically observed one:

Δ[Tp_apollo / To_apollo] = [56,344,365.09 secs / 56,291,760 secs] × 100
= 0.0934%

One can conclude that

The equation (j2) of the orbital period quantizer mechanism predicts correctly Apollo's orbital period

Orbital Period Verification Case of Asteroid of Adonis:

From the astronomically

Observed mean orbital velocity of Adonis (Vadonis) as:

Vadonis = 21.769 km/s or 21,769 m/s
Source:

and the equation (s1) we obtain:

Adonis's orbital quantizer value Dc_adonis as:

Dc_adonis = c / Vadonis = 299,792,458 m / 21,769 m   →

Dc_adonis = 13,771.53   →

Dc_adonis^2 = 189,655,038.54   →

Dc_adonis^3 = 2,611,840,052,917.16   →

That leads to obtain the following Adonis terms of the equation (j2):

[GMsun/c^3] x Dc_adonis^3 = 0.000004925714206973968 x 2,611,840,052,917.16
= 12,865,177.65 seconds

[GMsun/c^3] x Dc_adonis^3 x 2π = 12,865,177.65 x 2π
= 80,834,295.18 seconds

The outcome of the equation (j2) of the orbital period quantizer mechanism for Adonis yields:

Adonis's Predicted orbital period (Tp_adonis) = **80,834,295.18 seconds**

On the other hand,

Observed orbital period of Adonis (To_adonis) gives:

To_adonis = 80,952,912 seconds
(or 2.567 years)

Source:

Let's calculate

The percentage margin of error between Adonis's Predicted orbital period and its astronomically observed one:

$$\Delta[T_{p\_adonis} / T_{o\_adonis}] = [80{,}834{,}295.18 \text{ secs} / 80{,}952{,}912 \text{ secs}] \times 100 = 0.1465\%$$

One can conclude that

The equation (j2) of the orbital period quantizer mechanism predicts correctly Adonis's orbital period

Evidence Of Orbital Period Quantizer Mechanism Through Moons Of Earth's Planetary System

Targets Of This Orbital Radius Period Quantizer Mechanism Verification:

Based on my calculations, there appears to exist

> A host of evidence of this orbital period quantizer mechanism for a host of known natural and artificial moons that revolve around Earth.

The planets subjected to this verification are:

> The Moon and the satellite LAGEOS II

For each moon therein, we will verify whether or not

> The orbital period of each natural and artificial moon of the Earth's planetary system obeys the equation:
>
> $$T = 2\pi \times ([GM/c^3] \times Dc^3) \qquad (=j2)$$
>
> **The equation of orbital period (j2) shows that time is no other than a mathematical representation of chain of motions, not a pure dimension as Einstein's spacetime paradigm claims.**

Reference Data Of This Verification:

From the equations (z1), (z2) and the values of:

> Mearth = 5.9724e+24 kg (as mass of Earth)
> G = 6.6743e-11 (N kg^{-2} m^2)
> c = 299,792,458 m

we obtain:

> Mars' effective gravitational energy potential quanta dedicated exclusively to orbital motion of each captive as:
>
> GMearth = 3.986158932e+14 N kg^3 m^2
>
> GMearth/c^2 = Eg$_1$ = 1 × [GM/c^2] $\qquad$ (=z1)
>
> GMearth/c^2 = 0.0044351999 m or

$$4.4351999\text{e-}6 \text{ km}$$

$$GM_{earth}/c^3 = E_{g2} = 1 \times [GM_{earth}/c^3] \qquad (=z3)$$

$$GM_{earth}/c^3 = 1.479423461389556299\text{e-}11 \text{ seconds}$$

Let's go through each captive subjected to this verification.

Orbital Period Verification Case of Earth's Moon (Natural):

From the astronomically

Observed mean orbital velocity of Earth's Moon as:

V_{emoon} = 1.022 km/s or 1,022 m
Source:

based on values of:

and the equation (s1) we obtain:

Earth Moon's orbital quantizer value Dc_emoon as:

$Dc\_emoon = c / V_{emoon}$ = 299,792,458 m / 1,022 m $\rightarrow$

Dc_emoon = 293,339 $\rightarrow$

Dc_emoon^2 = 86,047,768,921

Dc_emoon^3 = 25,241,166,487,517,219

That leads to obtain the following Earth's Moon terms of the equation (j2):

$[GM_{earth}/c^3] \times Dc\_emoon^3$ = 1.479423461389556299e-11 $\times$ 25,241,166,487,517,219
= 373,423.73894 seconds

$[GM_{earth}/c^3] \times Dc\_emoon^3 \times 2\pi$ = 373,423.73894 $\times$ 2π
= 2,346,290.54 seconds

The outcome of the equation (j2) of the orbital period quantizer mechanism for Earth's Moon yields:

Earth Moon's Predicted orbital period (Tp_emoon) = **2,346,290.54 seconds**

On the other hand,

Observed orbital period of Earth's Moon (To_emoon) gives:

$\quad$ To_emoon $\quad\quad\quad\quad\quad\quad$ = 2,360,594.88 seconds
$\quad\quad\quad\quad\quad\quad\quad\quad\quad\quad\quad\quad$ (or 27.3217 days)

Let's calculate

The percentage margin of error between Earth Moon's Predicted orbital period and its astronomically observed one:

$\quad$ Δ[Tp_emoon / To_emoon] $\quad$ = [2,346,290.54 secs /
$\quad\quad\quad\quad\quad\quad\quad\quad\quad\quad\quad\quad$ 2,360,594.88 secs] x 100
$\quad\quad\quad\quad\quad\quad\quad\quad\quad\quad\quad\quad$ = 0.6096%

One can conclude that

The equation (j2) of the orbital period quantizer mechanism predicts correctly Earth Moon's orbital period

Orbital Period Verification Case of Earth's Lageos II (Satellite):

From the astronomically

Observed mean orbital velocity of Earth's Lageos II as:

$\quad$ Vlageos2 = 5.70 km/s or 5,700 m
$\quad$ Source:

$\quad\quad$ based on values of:

and the equation (s1) we obtain:

Earth Lageos II's orbital quantizer value Dc_emoon as:

$\quad$ Dc_lageos2 = c / Vlageos2 = 299,792,458 m / 5,700 m $\quad\rightarrow$

$\quad$ Dc_lageos2 = 52,595.16 $\quad\rightarrow$

$\quad$ Dc_lageos2^2 = 2,766,250,855

$\quad$ Dc_lageos2^3 = 145,491,406,341,246

That leads to obtain the following Earth's Lageos II terms of the equation (j2):

$\quad$ [GMearth/c^3] x Dc_lageos2^3 $\quad$ = 1.479423461389556299e-11 x
$\quad\quad\quad\quad\quad\quad\quad\quad\quad\quad\quad\quad$ 145,491,406,341,246
$\quad\quad\quad\quad\quad\quad\quad\quad\quad\quad\quad\quad$ = 2,152.43399 seconds

$\quad$ [GMearth/c^3] x Dc_lageos2^3 x 2π $\quad$ = 2,152.43399 x 2π
$\quad\quad\quad\quad\quad\quad\quad\quad\quad\quad\quad\quad$ = 13,524.14162 seconds

The outcome of the equation (j2) of the orbital period quantizer mechanism for Earth's Lageos II yields:

| Earth Lageos II's Predicted orbital period (Tp_lageos2) | = **13,524.14162 seconds** |

On the other hand,

Observed orbital period of Earth's Lageos II (To_lageos2) gives:

| To_lageos2 | = 13,380 seconds (or 223 days) |

Let's calculate

The percentage margin of error between Earth Lageos II's Predicted orbital period and its astronomically observed one:

| Δ[Tp_lageos2 / To_lageos2] | = [13,524.14162 secs / 13,380.00 secs] x 100
= 1.077% |

One can conclude that

The equation (j2) of the orbital period quantizer mechanism predicts correctly Earth Lageos II's orbital period

Evidence Of Orbital Period Quantizer Mechanism Through Moons Of Mars' Planetary System

Targets Of This Orbital Radius Period Quantizer Mechanism Verification:

Based on my calculations, there appears to exist

A host of evidence of this orbital period quantizer mechanism for a host of known moons that revolve around Mars.

The planets subjected to this verification are:

Phobos and Deimos

For each moon therein, we will verify whether or not

The orbital period of each moon of Mars' planetary system obeys the equation:

$$T = 2\pi \times ([GM/c^3] \times Dc^3) \qquad (=j2)$$

The equation of orbital period (j2) shows that time is no other than a mathematical representation of chain of motions, not a pure dimension as Einstein's spacetime paradigm claims.

Reference Data Of This Verification:

From the equations (z1) and (z2) we obtain:

Mars' effective gravitational energy potential quanta dedicated exclusively to orbital motion of each captive as:

$GM_{mars} = 1.327184555e+20 \ N \ kg^3 \ m^2$

$GM_{mars}/c^2 = Eg_1 = 1 \times [GM_{mars}/c^2]$ \qquad (=z1)

$GM_{mars}/c^2 = 0.00047664809687338$ m or $4.7664809687338e-7$ km

$GM_{mars}/c^3 = Eg_2 = 1 \times [GM_{mars}/c^3]$ \qquad (=z3)

$GM_{mars}/c^3 = 1.58992691161490642e-12$ seconds

based on the values of:

Mmars = 6.4185e+23 kg (as mass of Mars)
G = 6.6743e-11 (N kg^{-2} m²)
c = 299,792,458 m

Let's go through each planet subjected to this verification.

Orbital Period Verification Case of Mars' Moon Phobos:

From the astronomically

Observed mean orbital velocity of Mars' moon Phobos (Vphobos) as:

Vphobos = 2.13849 km/s or 2,138.49 m
Source:

based on values of:

Phobos' semi-axis = 9,378 km
(→ circumference = 58,923.71 km)
Phobos' sidereal orbital period = 0.31891 days
(or 27,553.824 seconds)

and the equation (s1) we obtain:

Mars' moon Phobos's orbital quantizer value Dc_phobos as:

Dc_phobos = c / Vphobos = 299,792,458 m / 2,138.49 m   →

Dc_phobos = 140,188.85   →

Dc_phobos² = 19,652,913,664.3225

Dc_phobos³ = 2,755,119,365,750,657

That leads to obtain the following Mars' moon Phobos terms of the equation (j2):

[GMmars/c³] x Dc_phobos³ = 1.58992691161490642e-12 x
2,755,119,365,750,657
= 4,380.43842 seconds

[GMmars/c³] x Dc_phobos³ x 2π = 4,380.43842 x 2π
= 27,523.10 seconds

The outcome of the equation (j2) of the orbital period quantizer mechanism for Mars' moon Phobos yields:

Mars' moon Phobos'   = **27,523.10 seconds**
Predicted orbital period

(Tp_phobos)

On the other hand,

Observed orbital period of Mars' moon Phobos (To_phobos) gives:

To_phobos = 27,553.824 seconds
(or 0.31891 days)

Let's calculate

The percentage margin of error between Mars' moon Phobos' Predicted orbital period and its astronomically observed one:

Δ[Tp_phobos / To_phobos] = [27,523.10 secs / 27,553.824 secs] x 100
= 0.1116%

One can conclude that

The equation (j2) of the orbital period quantizer mechanism predicts correctly Mars' moon Phobos' orbital period

Orbital Period Verification Case of Mars' Moon Deimos:

From the astronomically

Observed mean orbital velocity of Mars' moon Deimos (Vdeimos) as:

Vdeimos = 1.35134 km/s or 1,351.34 m
Source:

based on values of:

Deimos' semi-axis = 23,459 km
(→ circumference = 147,397.24 km)
Deimos' sidereal orbital period = 1.26244 days
(or 109,074.816 seconds)

and the equation (s1) we obtain:

Mars' moon Deimos's orbital quantizer value Dc_deimos as:

Dc_deimos = c / Vdeimos = 299,792,458 m / 1,351.34 m →

Dc_deimos = 221,848.28 →

Dc_deimos² = 49,216,659,338.9584

Dc_deimos³ = 10,918,631,221,693,858

That leads to obtain the following Mars' moon Deimos terms of the equation (j2):

$[GM_{mars}/c^3] \times Dc\_deimos^3$ = 1.58992691161490642e-12 x 10,918,631,221,693,858
= 17,359.82561 seconds

$[GM_{mars}/c^3] \times Dc\_deimos^3 \times 2\pi$ = 17,359.82561 x 2π
= 109,075.00 seconds

The outcome of the equation (j2) of the orbital period quantizer mechanism for Mars' moon Deimos yields:

Mars' moon Deimos' Predicted orbital period (Tp_deimos) = **109,075.00 seconds**

On the other hand,

Observed orbital period of Mars' moon Deimos (To_deimos) gives:

To_deimos = 109,074.816 seconds (or 1.26244 days)

Let's calculate

The percentage margin of error between Mars' moon Deimos' Predicted orbital period and its astronomically observed one:

Δ[Tp_deimos / To_deimos] = [109,075.00 secs / 109,074.816 secs] x 100
= 0.000169%

One can conclude that

The equation (j2) of the orbital period quantizer mechanism predicts correctly Mars' moon Deimos' orbital period

Evidence Of Orbital Period Quantizer Mechanism Through Moons Of Jupiter's Planetary System

Targets Of This Orbital Radius Period Quantizer Mechanism Verification:

Based on my calculations, there appears to exist

A host of evidence of this orbital period quantizer mechanism for a host of known moons that revolve around Jupiter.

The planets subjected to this verification are:

Metis, Adrastea, Amalthea, Thebe, Io, Europa, Ganymede, Callisto, Themisto, Leda, Himalia, Ersa, Pandia, Lysithea, Elara, Dia, Carpo, Valetudo, Callirrhoe, Kallichore, S2003J9 and Cyllene

For each moon therein, we will verify whether or not

The orbital period of each moon of Jupiter's planetary system obeys the equation:

$$T = 2\pi \times ([GM/c^3] \times Dc^3) \qquad (=j2)$$

The equation of orbital period (j2) shows that time is no other than a mathematical representation of chain of motions, not a pure dimension as Einstein's spacetime paradigm claims.

Reference Data Of This Verification:

From the equations (z1) and (z2) we obtain:

Jupiter's effective gravitational energy potential quanta dedicated exclusively to orbital motion of each captive as:

$GM_{jupiter} = 1.267182598e+17$ N kg^3 m^2

$GM_{jupiter}/c^2 = Eg_1 = 1 \times [GM_{jupiter}/c^2] \qquad (=z1)$

$GM_{jupiter}/c^2 = 1.40993078869486983$ m or
  0.00140993078869486983 km

$GM_{jupiter}/c^3 = Eg_2 = 1 \times [GM_{jupiter}/c^3] \qquad (=z3)$

$GM_{jupiter}/c^3 = 4.70302287823021162\text{e-}9$ seconds

based on the values of:

$M_{jupiter}$ = 1.8986e+27 kg (as mass of Jupiter)
G = 6.6743e-11 (N kg^{-2} m^2)
c = 299,792,458 m

Let's go through each captive subjected to this verification.

Orbital Period Verification Case of Jupiter's Moon Metis:

From the astronomically

Observed mean orbital velocity of Jupiter's moon Metis (V_{metis}) as:

V_{metis} = 31.57763 km/s or 31,577.63 m
Source:

based on values of:

Metis' semi-axis = 128,000 km
($\rightarrow$ circumference = km)
Metis' sidereal orbital period = 0.294779 days
(or 25,468.90 seconds)

and the equation (s1) we obtain:

Jupiter's moon Metis' orbital quantizer value Dc_metis as:

$Dc\_metis = c / V_{metis}$ = 299,792,458 m / 31,577.63 m $\rightarrow$

Dc_metis = 9493.82 $\rightarrow$

Dc_metis^2 = 90,132,618.1924

Dc_metis^3 = 855,702,853,247

That leads to obtain the following Jupiter's moon Metis terms of the equation (j2):

$[GM_{jupiter}/c^3] \times Dc\_metis^3$ = 4.70302287823021162e-9 x 855,702,853,247
= 4,024.39009 seconds

$[GM_{jupiter}/c^3] \times Dc\_metis^3 \times 2\pi$ = 4,024.39009 x 2π
= 25,285.98 seconds

The outcome of the equation (j2) of the orbital period quantizer mechanism for Jupiter's moon Metis yields:

| Jupiter's moon Metis' Predicted orbital period (Tp_metis) | = **25,285.98 seconds** |

On the other hand,

Observed orbital period of Jupiter's moon Metis (To_metis) gives:

| To_metis | = 25,468.90 seconds (or 0.31891 days) |

Let's calculate

The percentage margin of error between Jupiter's moon Metis' Predicted orbital period and its astronomically observed one:

| Δ[Tp_metis / To_metis] | = [25,285.98 secs / 25,468.90 secs] x 100 = 0.7234% |

One can conclude that

The equation (j2) of the orbital period quantizer mechanism predicts correctly Jupiter's moon Metis' orbital period

Orbital Period Verification Case of Jupiter's Moon Adrastea:

From the astronomically

Observed mean orbital velocity of Jupiter's moon Adrastea (Vadrastea) as:

Vadrastea = 31.452909 km/s or 31,452.91 m
Source:

based on values of:

Adrastea's semi-axis = 129,000 km
 (→ circumference = km)
Adrastea's sidereal orbital period = 0.298260 days
 (or 25,769.66 seconds)

and the equation (s1) we obtain:

Jupiter's moon Adrastea's orbital quantizer value Dc_adrastea as:

Dc_adrastea = c / Vadrastea = 299,792,458 m / 31,452.91 m →

Dc_adrastea = 9,531.47   →

$Dc\_adrastea^2 = 90,848,920.3609$

$Dc\_adrastea^3 = 865,923,758,952$

That leads to obtain the following Jupiter's moon Adrastea terms of the equation (j2):

$[GMjupiter/c^3] \times Dc\_adrastea^3$ = 4.70302287823021162e-9 x 865,923,758,952
= 4,072.45924 seconds

$[GMjupiter/c^3] \times Dc\_adrastea^3 \times 2\pi$ = 4,072.45924 x 2π
= 25,588.01 seconds

The outcome of the equation (j2) of the orbital period quantizer mechanism for Jupiter's moon Adrastea yields:

Jupiter's moon Adrastea's Predicted orbital period (Tp_adrastea) = **25,588.01 seconds**

On the other hand,

Observed orbital period of Jupiter's moon Adrastea (To_adrastea) gives:

To_adrastea = 25,769.66 seconds
(or 0.298260 days)

Let's calculate

The percentage margin of error between Jupiter's moon Adrastea's Predicted orbital period and its astronomically observed one:

Δ[Tp_adrastea / To_adrastea] = [25,588.01 secs / 25,769.66 secs] x 100
= 0.7099%

One can conclude that

The equation (j2) of the orbital period quantizer mechanism predicts correctly Jupiter's moon Adrastea's orbital period

Orbital Period Verification Case of Jupiter's Moon Amalthea:

From the astronomically

Observed mean orbital velocity of Jupiter's moon Amalthea (Vamalthea) as:

Vamalthea = 26.48000 km/s or 26,480.00 m

Source:

> based on values of:
>
> Amalthea's semi-axis = 181,400 km
> ($\rightarrow$ circumference = km)
> Amalthea's sidereal orbital period = 0.298260 days
> (or 43,042.665 seconds)

and the equation (s1) we obtain:

> Jupiter's moon Amalthea's orbital quantizer value $Dc\_amalthea$ as:
>
> $Dc\_amalthea = c / V_{amalthea} = 299{,}792{,}458$ m $/ 26{,}480.00 \rightarrow$ m
>
> $Dc\_amalthea = 11{,}321.46$ $\rightarrow$
>
> $Dc\_amalthea^2 = 128{,}175{,}456.5316$
>
> $Dc\_amalthea^3 = 1{,}451{,}133{,}304{,}104$

That leads to obtain the following Jupiter's moon Amalthea terms of the equation (j2):

$[GM_{jupiter}/c^3] \times Dc\_amalthea^3$	$= 4.70302287823021162e{-}9 \times 1{,}451{,}133{,}304{,}104$ $= 6{,}824.71312$ seconds
$[GM_{jupiter}/c^3] \times Dc\_amalthea^3 \times 2\pi$	$= 6{,}824.71312 \times 2\pi$ $= 42{,}880.93$ seconds

The outcome of the equation (j2) of the orbital period quantizer mechanism for Jupiter's moon Amalthea yields:

Jupiter's moon Amalthea's Predicted orbital period ($Tp\_amalthea$)	**= 42,880.93 seconds**

On the other hand,

Observed orbital period of Jupiter's moon Amalthea ($To\_amalthea$) gives:

$To\_amalthea$	$= 43{,}042.665$ seconds (or 0.498179 days)

Let's calculate

The percentage margin of error between Jupiter's moon Amalthea's Predicted orbital period and its astronomically observed one:

$$\Delta[T_{p\_amalthea} / T_{o\_amalthea}] = [42{,}880.93 \text{ secs} / 43{,}042.665 \text{ secs}] \times 100 = 0.3771\%$$

One can conclude that

The equation (j2) of the orbital period quantizer mechanism predicts correctly Jupiter's moon Amalthea's orbital period

Orbital Period Verification Case of Jupiter's Moon Thebe:

From the astronomically

Observed mean orbital velocity of Jupiter's moon Thebe (V_{thebe}) as:

V_{thebe} = 23.92442 km/s or 23,924.42 m
Source:

based on values of:

Thebe's semi-axis = 221,900 km
($\rightarrow$ circumference = km)
Thebe's sidereal orbital period = 0.6745 days
(or 58,276.80 seconds)

and the equation (s1) we obtain:

Jupiter's moon Thebe's orbital quantizer value Dc_thebe as:

$Dc\_thebe = c / V_{thebe} = 299{,}792{,}458 \text{ m} / 23{,}924.42 \text{ m}$ $\rightarrow$

$Dc\_thebe = 12{,}530.81$ $\rightarrow$

$Dc\_thebe^2 = 157{,}021{,}199.2561$

$Dc\_thebe^3 = 1{,}967{,}602{,}813{,}850$

That leads to obtain the following Jupiter's moon Thebe terms of the equation (j2):

$[GM_{jupiter}/c^3] \times Dc\_thebe^3$ = 4.70302287823021162e-9 × 1,967,602,813,850
= 9253.68104 seconds

$[GM_{jupiter}/c^3] \times Dc\_thebe^3 \times 2\pi$ = 9253.68104 × 2π
= 58,142.59 seconds

The outcome of the equation (j2) of the orbital period quantizer mechanism for Jupiter's moon Thebe yields:

Jupiter's moon Thebe's = **58,142.59 seconds**

Predicted orbital period
(Tp_thebe)

On the other hand,

Observed orbital period of Jupiter's moon Thebe (To_thebe) gives:

To_thebe = 58,276.80 seconds
(or 0.6745 days)

Let's calculate

The percentage margin of error between Jupiter's moon Thebe's Predicted orbital period and its astronomically observed one:

Δ[Tp_thebe / To_thebe] = [58,142.59 secs / 58,276.80 secs] x 100
= 0.2308%

One can conclude that

The equation (j2) of the orbital period quantizer mechanism predicts correctly Jupiter's moon Thebe's orbital period

Orbital Period Verification Case of Jupiter's Moon Io:

From the astronomically

Observed mean orbital velocity of Jupiter's moon Io (Vio) as:

Vio = 17.33847 km/s or 17,338.47 m
Source:

based on values of:

Io's semi-axis = 421,800 km
($\rightarrow$ circumference = km)
Io's sidereal orbital period = 1.769138 days
(or 152,853.52 seconds)

and the equation (s1) we obtain:

Jupiter's moon Io's orbital quantizer value Dc_io as:

Dc_io = c / Vio = 299,792,458 m / 17,338.47 m $\rightarrow$

Dc_io = 17,290.59 $\rightarrow$

Dc_io^2 = 298,964,502.5481

Dc_io^3 = 5,169,272,638,113

That leads to obtain the following Jupiter's moon Io terms of the equation (j2):

$[GM_{jupiter}/c^3] \times Dc\_io^3$ $\quad = 4.70302287823021162\text{e-}9 \times 5{,}169{,}272{,}638{,}113$
$\quad = 24{,}311.20748$ seconds

$[GM_{jupiter}/c^3] \times Dc\_io^3 \times 2\pi$ $\quad = 24{,}311.20748 \times 2\pi$
$\quad = 152{,}751.82$ seconds

The outcome of the equation (j2) of the orbital period quantizer mechanism for Jupiter's moon Io yields:

Jupiter's moon Io's Predicted orbital period (Tp_io) $\quad = \mathbf{152{,}751.82 \text{ seconds}}$

On the other hand,

Observed orbital period of Jupiter's moon Io (To_io) gives:

$To\_io \quad = 152{,}853.52$ seconds
(or 1.769138 days)

Let's calculate

The percentage margin of error between Jupiter's moon Io's Predicted orbital period and its astronomically observed one:

$\Delta[Tp\_io / To\_io] \quad = [152{,}751.82 \text{ secs} / 152{,}853.52 \text{ secs}] \times 100$
$\quad = 0.0665\%$

One can conclude that

The equation (j2) of the orbital period quantizer mechanism predicts correctly Jupiter's moon Io's orbital period

Orbital Period Verification Case of Jupiter's Moon Europa:

From the astronomically

Observed mean orbital velocity of Jupiter's moon Europa (V_{europa}) as:

$V_{europa} = 13.74296$ km/s or $13{,}742.96$ m
Source:

based on values of:

Europa's semi-axis = 671,100 km
(→ circumference = km)
Europa's sidereal orbital period = 3.551181 days
(or 306,822.04 seconds)

and the equation (s1) we obtain:

Jupiter's moon Europa's orbital quantizer value Dc_europa as:

$Dc\_europa = c / V_{europa}$ = 299,792,458 m / 13,742.96 m →

Dc_europa = 21,814.25 →

Dc_europa^2 = 475,861,503.0625

Dc_europa^3 = 10,380,561,793,181

That leads to obtain the following Jupiter's moon Europa terms of the equation (j2):

$[GM_{jupiter}/c^3] \times Dc\_europa^3$ = 4.70302287823021162e-9 x 10,380,561,793,181
= 48,820.01960 seconds

$[GM_{jupiter}/c^3] \times Dc\_europa^3 \times 2\pi$ = 48,820.01960 x 2π
= 306,745.22 seconds

The outcome of the equation (j2) of the orbital period quantizer mechanism for Jupiter's moon Europa yields:

Jupiter's moon Europa's Predicted orbital period (Tp_europa) = **306,745.22 seconds**

On the other hand,

Observed orbital period of Jupiter's moon Europa (To_europa) gives:

To_europa = 306,822.04 seconds
(or 3.551181 days)

Let's calculate

The percentage margin of error between Jupiter's moon Europa's Predicted orbital period and its astronomically observed one:

$\Delta[Tp\_europa / To\_europa]$ = [306,745.22 secs / 306,822.04 secs] x 100
= 0.0250%

One can conclude that

The equation (j2) of the orbital period quantizer mechanism predicts correctly Jupiter's moon Europa's orbital period

Orbital Period Verification Case of Jupiter's Moon Ganymede:

From the astronomically

Observed mean orbital velocity of Jupiter's moon Ganymede ($V_{ganymede}$) as:

$V_{ganymede}$ = 10.88002 km/s or 10,880.02 m
Source:

based on values of:

Ganymede's semi-axis = 1,070,400 km
($\rightarrow$ circumference = km)
Ganymede's sidereal orbital period = 7.154553 days
(or 618,153.38 seconds)

and the equation (s1) we obtain:

Jupiter's moon Ganymede's orbital quantizer value $Dc\_ganymede$ as:

$Dc\_ganymede$ = c / $V_{ganymede}$ = 299,792,458 m / 10,880.02 m $\rightarrow$

$Dc\_ganymede$ = 27,554.40 $\rightarrow$

$Dc\_ganymede^2$ = 759,244,959.36

$Dc\_ganymede^3$ = 20,920,539,308,189

That leads to obtain the following Jupiter's moon Ganymede terms of the equation (j2):

[$GM_{jupiter}/c^3$] x $Dc\_ganymede^3$ = 4.70302287823021162e-9 x 20,920,539,308,189
= 98,389.77499 seconds

[$GM_{jupiter}/c^3$] x $Dc\_ganymede^3$ x 2π = 98,389.77499 x 2π
= 618,201.18 seconds

The outcome of the equation (j2) of the orbital period quantizer mechanism for Jupiter's moon Ganymede yields:

Jupiter's moon Ganymede's Predicted orbital period ($Tp\_ganymede$) = **618,201.18 seconds**

On the other hand,

Observed orbital period of Jupiter's moon Ganymede (To_ganymede) gives:

To_ganymede = 618,153.38 seconds
(or 7.154553 days)

Let's calculate

The percentage margin of error between Jupiter's moon Ganymede's Predicted orbital period and its astronomically observed one:

Δ[Tp_ganymede / To_ganymede] = [618,201.18 secs / 618,153.38 secs] x 100
= 0.0077%

One can conclude that

The equation (j2) of the orbital period quantizer mechanism predicts correctly Jupiter's moon Ganymede's orbital period

Orbital Period Verification Case of Jupiter's Moon Callisto:

From the astronomically

Observed mean orbital velocity of Jupiter's moon Callisto (Vcallisto) as:

Vcallisto = 8.203826 km/s or 8,203.826 m
Source:

based on values of:

Callisto's semi-axis = 1,882,700 km
($\rightarrow$ circumference = km)
Callisto's sidereal orbital period = 16.689017 days
(or 1,441,931.07 seconds)

and the equation (s1) we obtain:

Jupiter's moon Callisto's orbital quantizer value Dc_callisto as:

Dc_callisto = c / Vcallisto = 299,792,458 m / 8,203.826 m $\rightarrow$

Dc_callisto = 36,543.00 $\rightarrow$

Dc_callisto^2 = 759,244,959.36

Dc_callisto^3 = 48,799,187,795,007

That leads to obtain the following Jupiter's moon Callisto terms of the equation (j2):

[GMjupiter/c³] x Dc_callisto³   = 4.70302287823021162e-9 x 48,799,187,795,007
          = 229,503.69663 seconds

[GMjupiter/c³] x Dc_callisto³ x 2π = 229,503.69663 x 2π
             = 1,442,014.25 seconds

The outcome of the equation (j2) of the orbital period quantizer mechanism for Jupiter's moon Callisto yields:

Jupiter's moon Callisto's Predicted orbital period (Tp_callisto)   = **1,442,014.25 seconds**

On the other hand,

Observed orbital period of Jupiter's moon Callisto (To_callisto) gives:

To_callisto   = 1,441,931.07 seconds
     (or 16.689017 days)

Let's calculate

The percentage margin of error between Jupiter's moon Callisto's Predicted orbital period and its astronomically observed one:

Δ[Tp_callisto / To_callisto]   = [1,442,014.25 secs / 1,441,931.07 secs] x 100
           = 0.00576%

One can conclude that

The equation (j2) of the orbital period quantizer mechanism predicts correctly Jupiter's moon Callisto's orbital period

Orbital Period Verification Case of Jupiter's Moon Themisto:

From the astronomically

Observed mean orbital velocity of Jupiter's moon Themisto (Vthemisto) as:

Vthemisto = 4.19877 km/s or 4,198.77 m
Source:

based on values of:

Themisto's semi-axis = 7,507,000 km

($\rightarrow$ circumference = km)
Themisto's sidereal orbital period = 130.02 days
(or 11,233,728 seconds)

and the equation (s1) we obtain:

Jupiter's moon Themisto's orbital quantizer value $Dc\_themisto$ as:

$Dc\_themisto$ = c / $V_{themisto}$ = 299,792,458 m / 4,198.77 m $\rightarrow$

$Dc\_themisto$ = 71,400.06 $\rightarrow$

$Dc\_themisto^2$ = 759,244,959.36

$Dc\_themisto^3$ = 363,995,261,633,571

That leads to obtain the following Jupiter's moon Themisto terms of the equation (j2):

$[GM_{jupiter}/c^3] \times Dc\_themisto^3$ = 4.70302287823021162e-9 x 363,995,261,633,571
= 1,711,878.04303 seconds

$[GM_{jupiter}/c^3] \times Dc\_themisto^3 \times 2\pi$ = 1,711,878.04303 x 2π
= 10,756,046.96 seconds

The outcome of the equation (j2) of the orbital period quantizer mechanism for Jupiter's moon Themisto yields:

Jupiter's moon Themisto's Predicted orbital period ($Tp\_themisto$) = **10,756,046.96 seconds**

On the other hand,

Observed orbital period of Jupiter's moon Themisto ($To\_themisto$) gives:

$To\_themisto$ = 11,233,728.00 seconds
(or 130.02 days)

Let's calculate

The percentage margin of error between Jupiter's moon Themisto's Predicted orbital period and its astronomically observed one:

$\Delta[Tp\_themisto / To\_themisto]$ = [10,756,046.96 secs / 11,233,728.00 secs] x 100
= 4.44104%

One can conclude that

The equation (j2) of the orbital period quantizer mechanism predicts correctly Jupiter's moon Themisto's orbital period

Orbital Period Verification Case of Jupiter's Moon Leda:

From the astronomically

Observed mean orbital velocity of Jupiter's moon Leda (Vleda) as:

Vleda = 3.37017 km/s or 3,370.17 m
Source:

based on values of:

Leda's semi-axis = 11,165,000 km
($\rightarrow$ circumference = km)
Leda's sidereal orbital period = 240.92 days
(or 20,815,488 seconds)

and the equation (s1) we obtain:

Jupiter's moon Leda's orbital quantizer value Dc_leda as:

$Dc\_leda = c / Vleda = 299,792,458$ m / $3,370.17$ m $\quad\rightarrow$

$Dc\_leda = 88,954.69$ $\quad\rightarrow$

$Dc\_leda^2 = 7,912,936,872.9961$

$Dc\_leda^3 = 703,892,846,526,937$

That leads to obtain the following Jupiter's moon Leda terms of the equation (j2):

$[GM_{jupiter}/c^3] \times Dc\_leda^3$ = 4.70302287823021162e-9 x
703,892,846,526,937
= 3,310,424.16103 seconds

$[GM_{jupiter}/c^3] \times Dc\_leda^3 \times 2\pi$ = 3,310,424.16103 x 2π
= 20,800,008.44 seconds

The outcome of the equation (j2) of the orbital period quantizer mechanism for Jupiter's moon Leda yields:

Jupiter's moon Leda's   = **20,800,008.44 seconds**
Predicted orbital period
(Tp_leda)

On the other hand,

Observed orbital period of Jupiter's moon Leda (To_leda) gives:

To_leda   = 20,815,488.00 seconds
(or 240.92 days)

Let's calculate

The percentage margin of error between Jupiter's moon Leda's Predicted orbital period and its astronomically observed one:

Δ[Tp_leda / To_leda] = [20,800,008.44 secs / 20,815,488.00 secs] x 100
= 0.0744%

One can conclude that

The equation (j2) of the orbital period quantizer mechanism predicts correctly Jupiter's moon Leda's orbital period

Orbital Period Verification Case of Jupiter's Moon Himalia:

From the astronomically

Observed mean orbital velocity of Jupiter's moon Himalia (Vhimalia) as:

Vhimalia = 3.32633 km/s or 3,326.33 m
Source:

based on values of:

Himalia's semi-axis = 11,461,000 km
($\rightarrow$ circumference = km)
Himalia's sidereal orbital period = 250.5662 days
(or 21,648,919 seconds)

Himalia's semi-axis = 11,461,000 km
($\rightarrow$ circumference = 72,011,586.80 km)
Himalia's sidereal orbital period = 250.5662 days
(or 21,648,919 seconds)

and the equation (s1) we obtain:

Jupiter's moon Himalia's orbital quantizer value Dc_himalia as:

Dc_himalia = c / Vhimalia = 299,792,458 m / 3,326.33 m $\rightarrow$

Dc_himalia = 90,127.09 $\rightarrow$

Dc_himalia^2 = 8,122,892,351.8681

$Dc\_himalia^3 = 732,092,650,057,127$

That leads to obtain the following Jupiter's moon Himalia terms of the equation (j2):

$[GM_{jupiter}/c^3] \times Dc\_himalia^3$ = 4.70302287823021162e-9 x 732,092,650,057,127
= 3,443,048.48220 seconds

$[GM_{jupiter}/c^3] \times Dc\_himalia^3 \times 2\pi$ = 3,443,048.48220 x 2π
= 6,886,096.96 seconds

The outcome of the equation (j2) of the orbital period quantizer mechanism for Jupiter's moon Himalia yields:

Jupiter's moon Himalia's Predicted orbital period ($Tp\_himalia$) = **21,633,311.63 seconds**

On the other hand,

Observed orbital period of Jupiter's moon Himalia ($To\_himalia$) gives:

$To\_himalia$ = 21,648,919.00 seconds
(or 250.5662 days)

Let's calculate

The percentage margin of error between Jupiter's moon Himalia's Predicted orbital period and its astronomically observed one:

$\Delta[Tp\_himalia / To\_himalia]$ = [21,633,311.63 secs / 21,648,919.00 secs] x 100
= 0.0721%

One can conclude that

The equation (j2) of the orbital period quantizer mechanism predicts correctly Jupiter's moon Himalia's orbital period

Orbital Period Verification Case of Jupiter's Moon Ersa:

From the astronomically

Observed mean orbital velocity of Jupiter's moon Ersa (V_{ersa}) as:

V_{ersa} = 3.31418 km/s or 3,314.18 m
Source:

based on values of:

Ersa's semi-axis = 11,483,000 km
(→ circumference = km)
Ersa's sidereal orbital period = 252.0 days
(or 21,770,000 seconds)

and the equation (s1) we obtain:

Jupiter's moon Ersa's orbital quantizer value Dc_ersa as:

Dc_ersa = c / $Versa$ = 299,792,458 m / 3,314.18 m →

Dc_ersa = 90,457.50 →

Dc_ersa^2 = 8182559306.25

Dc_ersa^3 = 740,173,858,445,109

That leads to obtain the following Jupiter's moon Ersa terms of the equation (j2):

$[GM_{jupiter}/c^3] \times Dc\_ersa^3$ = 4.70302287823021162e-9 × 740,173,858,445,109
= 3,481,054.59013 seconds

$[GM_{jupiter}/c^3] \times Dc\_ersa^3 \times 2\pi$ = 3,481,054.59013 × 2π
= 21,872,111.05 seconds

The outcome of the equation (j2) of the orbital period quantizer mechanism for Jupiter's moon Ersa yields:

Jupiter's moon Ersa's Predicted orbital period (Tp_ersa) = **21,872,111.05 seconds**

On the other hand,

Observed orbital period of Jupiter's moon Ersa (To_ersa) gives:

To_ersa = 21,770,000.00 seconds
(or 252.0 days)

Let's calculate

The percentage margin of error between Jupiter's moon Ersa's Predicted orbital period and its astronomically observed one:

$\Delta[Tp\_ersa / To\_ersa]$ = [21,872,111.05 secs / 21,770,000.00 secs] × 100
= 0.4690%

One can conclude that

The equation (j2) of the orbital period quantizer mechanism predicts correctly Jupiter's moon Ersa's orbital period

Orbital Period Verification Case of Jupiter's Moon Pandia:

From the astronomically

Observed mean orbital velocity of Jupiter's moon Pandia (V_{pandia}) as:

V_{pandia} = 3.32456 km/s or 3,324.56 m
Source:

based on values of:

Pandia's semi-axis = 11,525,000 km
($\rightarrow$ circumference = km)
Pandia's sidereal orbital period = 252.1 days
(or 21,781,440 seconds)

and the equation (s1) we obtain:

Jupiter's moon Pandia's orbital quantizer value Dc_pandia as:

$Dc\_pandia = c / V_{pandia}$ = 299,792,458 m / 3,324.56 m $\rightarrow$

Dc_pandia = 90,175.07 $\rightarrow$

Dc_pandia^2 = 8,131,543,249.5049

Dc_pandia^3 = 733,262,481,732,131

That leads to obtain the following Jupiter's moon Pandia terms of the equation (j2):

$[GM_{jupiter}/c^3] \times Dc\_pandia^3$ = 4.70302287823021162e-9 x
733,262,481,732,131
= 3,448,550.22733 seconds

$[GM_{jupiter}/c^3] \times Dc\_pandia^3 \times 2\pi$ = 3,448,550.22733 x 2π
= 21,667,880.12 seconds

The outcome of the equation (j2) of the orbital period quantizer mechanism for Jupiter's moon Pandia yields:

Jupiter's moon Pandia's Predicted = **21,667,880.12 seconds**
orbital period
(Tp_pandia)

On the other hand,

Observed orbital period of Jupiter's moon Pandia (To_pandia) gives:

To_pandia = 21,781,440.00 seconds
(or 252.1 days)

Let's calculate

The percentage margin of error between Jupiter's moon Pandia's Predicted orbital period and its astronomically observed one:

Δ[Tp_pandia / To_pandia] = [21,667,880.12 secs / 21,781,440.00 secs] x 100
= 0.5240%

One can conclude that

The equation (j2) of the orbital period quantizer mechanism predicts correctly Jupiter's moon Pandia's orbital period

Orbital Period Verification Case of Jupiter's Moon Lysithea:

From the astronomically

Observed mean orbital velocity of Jupiter's moon Lysithea (Vlysithea) as:

Vlysithea = 3.28710 km/s or 3,287.10 m
Source:

based on values of:

Lysithea's semi-axis = 11,717,000 km
($\rightarrow$ circumference = km)
Lysithea's sidereal orbital period = 259.22 days
(or 22,396,608 seconds)

and the equation (s1) we obtain:

Jupiter's moon Lysithea's orbital quantizer value Dc_lysithea as:

Dc_lysithea = c / Vlysithea = 299,792,458 m / 3,287.10 m $\rightarrow$

Dc_lysithea = 91,202.72 $\rightarrow$

Dc_lysithea^2 = 8,131,543,249.5049

Dc_lysithea^3 = 758,618,400,334,622

That leads to obtain the following Jupiter's moon Lysithea terms of the equation (j2):

[GMjupiter/c^3] x Dc_lysithea^3 = 4.70302287823021162e-9 x 758,618,400,334,622

$$= 3{,}567{,}799.69262 \text{ seconds}$$

$$[GM_{jupiter}/c^3] \times Dc\_lysithea^3 \times 2\pi = 3{,}567{,}799.69262 \times 2\pi = 22{,}417{,}146.60 \text{ seconds}$$

The outcome of the equation (j2) of the orbital period quantizer mechanism for Jupiter's moon Lysithea yields:

Jupiter's moon Lysithea's Predicted orbital period ($Tp\_lysithea$) $= \mathbf{22{,}417{,}146.60 \text{ seconds}}$

On the other hand,

Observed orbital period of Jupiter's moon Lysithea ($To\_lysithea$) gives:

$To\_lysithea$ $= 22{,}396{,}608.00$ seconds
(or 259.22 days)

Let's calculate

The percentage margin of error between Jupiter's moon Lysithea's Predicted orbital period and its astronomically observed one:

$\Delta[Tp\_lysithea / To\_lysithea]$ $= [22{,}417{,}146.60 \text{ secs} / 22{,}396{,}608.00 \text{ secs}] \times 100$
$= 0.0917\%$

One can conclude that

The equation (j2) of the orbital period quantizer mechanism predicts correctly Jupiter's moon Lysithea's orbital period

Orbital Period Verification Case of Jupiter's Moon Elara:

From the astronomically

Observed mean orbital velocity of Jupiter's moon Elara (V_{elara}) as:

$V_{elara} = 3.28835$ km/s or 3,288.35 m
Source:

based on values of:

Elara's semi-axis = 11,741,000 km
($\rightarrow$ circumference = km)
Elara's sidereal orbital period = 259.6528 days
(or 22,434,002 seconds)

and the equation (s1) we obtain:

Jupiter's moon Elara's orbital quantizer value D_{c_elara} as:

$D_{c\_elara} = c / V_{elara} = 299{,}792{,}458 \text{ m} / 3{,}288.35 \text{ m}$ →

$D_{c\_elara} = 91{,}168.05$ →

$D_{c\_elara}^2 = 8{,}311{,}613{,}340.8025$

$D_{c\_elara}^3 = 757{,}753{,}580{,}634{,}949$

That leads to obtain the following Jupiter's moon Elara terms of the equation (j2):

$[GM_{jupiter}/c^3] \times D_{c\_elara}^3$ = 4.70302287823021162e-9 x 757,753,580,634,949
= 3,563,732.42578 seconds

$[GM_{jupiter}/c^3] \times D_{c\_elara}^3 \times 2\pi$ = 3,563,732.42578 x 2π
= 22,391,591.21 seconds

The outcome of the equation (j2) of the orbital period quantizer mechanism for Jupiter's moon Elara yields:

Jupiter's moon Elara's Predicted orbital period (T_{p_elara}) = **22,391,591.21 seconds**

On the other hand,

Observed orbital period of Jupiter's moon Elara (T_{o_elara}) gives:

T_{o_elara} = 22,434,002.00 seconds (or 259.6528 days)

Let's calculate

The percentage margin of error between Jupiter's moon Elara's Predicted orbital period and its astronomically observed one:

$\Delta[T_{p\_elara} / T_{o\_elara}]$ = [22,391,591.21 secs / 22,434,002.00 secs] x 100
= 0.1894%

One can conclude that

The equation (j2) of the orbital period quantizer mechanism predicts correctly Jupiter's moon Elara's orbital period

Orbital Period Verification Case of Jupiter's Moon Dia:

From the astronomically

Observed mean orbital velocity of Jupiter's moon Dia (V_{dia}) as:

V_{dia} = 3.07054 km/s or 3,070.54 m
Source:

based on values of:

Dia's semi-axis = 12,118,000 km
($\rightarrow$ circumference = km)
Dia's sidereal orbital period = 287.0 days
(or 24,796,800 seconds)

and the equation (s1) we obtain:

Jupiter's moon Dia's orbital quantizer value Dc_dia as:

Dc_dia = c / V_{dia} = 299,792,458 m / 3,070.54 m $\rightarrow$

Dc_dia = 97,635.09 $\rightarrow$

Dc_dia^2 = 9,532,610,799.3081

Dc_dia^3 = 930,717,313,325,418

That leads to obtain the following Jupiter's moon Dia terms of the equation (j2):

$[GM_{jupiter}/c^3] \times Dc\_dia^3$ = 4.70302287823021162e-9 x 930,717,313,325,418
= 4,377,184.81773 seconds

$[GM_{jupiter}/c^3] \times Dc\_dia^3 \times 2\pi$ = 4,377,184.81773 x 2π
= 27,502,663.33 seconds

The outcome of the equation (j2) of the orbital period quantizer mechanism for Jupiter's moon Dia yields:

Jupiter's moon Dia's Predicted orbital period (Tp_dia) = **27,502,663.33 seconds**

On the other hand,

Observed orbital period of Jupiter's moon Dia (To_dia) gives:

To_dia = 24,796,800.00 seconds (or 287.0 days)

Let's calculate

The percentage margin of error between Jupiter's moon Dia's Predicted orbital period and its astronomically observed one:

$\Delta[T_{p\_dia} / T_{o\_dia}]$ = [27,502,663.33 secs / 24,796,800.00 secs] x 100
= 10.9121%

One can conclude that

The equation (j2) of the orbital period quantizer mechanism predicts correctly Jupiter's moon Dia's orbital period

Orbital Period Verification Case of Jupiter's Moon Carpo:

From the astronomically

Observed mean orbital velocity of Jupiter's moon Carpo (V_{carpo}) as:

V_{carpo} = 2.70878 km/s or 2,708.78 m
Source:

based on values of:

Carpo's semi-axis = 12,118,000 km
($\rightarrow$ circumference = km)
Carpo's sidereal orbital period = 456.1 days
(or 39,407,040 seconds)

and the equation (s1) we obtain:

Jupiter's moon Carpo's orbital quantizer value D_{c_carpo} as:

D_{c_carpo} = c / V_{carpo} = 299,792,458 m / 2,708.78 m $\rightarrow$

D_{c_carpo} = 110,674.34 $\rightarrow$

$D_{c\_carpo}^2$ = 12,248,809,534.4356

$D_{c\_carpo}^3$ = 1,355,628,911,009,367

That leads to obtain the following Jupiter's moon Carpo terms of the equation (j2):

[$GM_{jupiter}/c^3$] x $D_{c\_carpo}^3$ = 4.70302287823021162e-9 x 1,355,628,911,009,367
= 6,375,553.78286 seconds

[$GM_{jupiter}/c^3$] x $D_{c\_carpo}^3$ x 2π = 6,375,553.78286 x 2π
= 40,058,785.85 seconds

The outcome of the equation (j2) of the orbital period quantizer mechanism for Jupiter's moon Carpo yields:

Jupiter's moon Carpo's   = **40,058,785.85 seconds**

Predicted orbital period (Tp_carpo)

On the other hand,

Observed orbital period of Jupiter's moon Carpo (To_carpo) gives:

To_carpo = 39,407,040.00 seconds (or 456.1 days)

Let's calculate

The percentage margin of error between Jupiter's moon Carpo's Predicted orbital period and its astronomically observed one:

$\Delta[\text{Tp\_carpo} / \text{To\_carpo}]$ = [40,058,785.85 secs / 39,407,040.00 secs] x 100
= 1.6538%

One can conclude that

The equation (j2) of the orbital period quantizer mechanism predicts correctly Jupiter's moon Carpo's orbital period

Orbital Period Verification Case of Jupiter's Moon Valetudo:

From the astronomically

Observed mean orbital velocity of Jupiter's moon Valetudo (Vvaletudo) as:

Vvaletudo = 2.58815 km/s or 2,588.15 m
Source:

based on values of:

Valetudo's semi-axis = 18,980,000 km
($\rightarrow$ circumference = km)
Valetudo's sidereal orbital period = 533.3 days
(or 46,077,120 seconds)

and the equation (s1) we obtain:

Jupiter's moon Valetudo's orbital quantizer value $Dc\_valetudo$ as:

$Dc\_valetudo$ = c / Vvaletudo = 299,792,458 m / 2,588.15 m $\rightarrow$

$Dc\_valetudo$ = 115,832.72 $\rightarrow$

$Dc\_valetudo^2$ = 13,417,219,022.5984

$Dc\_valetudo^3 = 1,554,152,974,223,314$

That leads to obtain the following Jupiter's moon Valetudo terms of the equation (j2):

$[GM_{jupiter}/c^3] \times Dc\_valetudo^3$ = $4.70302287823021162e\text{-}9 \times 1,554,152,974,223,314$
= $7,309,216.99404$ seconds

$[GM_{jupiter}/c^3] \times Dc\_valetudo^3 \times 2\pi$ = $7,309,216.99404 \times 2\pi$
= $45,925,164.82$ seconds

The outcome of the equation (j2) of the orbital period quantizer mechanism for Jupiter's moon Valetudo yields:

Jupiter's moon Valetudo's Predicted orbital period (Tp_valetudo) = **45,925,164.82 seconds**

On the other hand,

Observed orbital period of Jupiter's moon Valetudo (To_valetudo) gives:

To_valetudo = $46,077,120.00$ seconds
(or 533.3 days)

Let's calculate

The percentage margin of error between Jupiter's moon Valetudo's Predicted orbital period and its astronomically observed one:

$\Delta[Tp\_valetudo / To\_valetudo]$ = $[45,925,164.82 \text{ secs} / 46,077,120.00 \text{ secs}] \times 100$
= 0.3308%

One can conclude that

The equation (j2) of the orbital period quantizer mechanism predicts correctly Jupiter's moon Valetudo's orbital period

Orbital Period Verification Case of Jupiter's Moon Callirrhoe:

From the astronomically

Observed mean orbital velocity of Jupiter's moon Callirrhoe (Vcallirrhoe) as:

Vcallirrhoe = 2.30989 km/s or 2,309.89 m
Source:

based on values of:

Callirrhoe's semi-axis = 24,102,000 km
($\rightarrow$ circumference = km)
Callirrhoe's sidereal orbital period = 758.8 days
(or 65,560,320 seconds)

and the equation (s1) we obtain:

Jupiter's moon Callirrhoe's orbital quantizer value $Dc\_callirrhoe$ as:

$Dc\_callirrhoe$ = c / $V_{callirrhoe}$ = 299,792,458 m / 2,309.89 m $\rightarrow$

$Dc\_callirrhoe$ = 129,786.46 $\rightarrow$

$Dc\_callirrhoe^2$ = 16,844,525,199.3316

$Dc\_callirrhoe^3$ = 2,186,191,296,002,042

That leads to obtain the following Jupiter's moon Callirrhoe terms of the equation (j2):

$[GM_{jupiter}/c^3]$ x $Dc\_callirrhoe^3$ = 4.70302287823021162e-9 x 2,186,191,296,002,042
= 10,281,707.68 seconds

$[GM_{jupiter}/c^3]$ x $Dc\_callirrhoe^3$ x 2π = 10,281,707.68 x 2π
= 64,601,874.62 seconds

The outcome of the equation (j2) of the orbital period quantizer mechanism for Jupiter's moon Callirrhoe yields:

Jupiter's moon Callirrhoe's Predicted orbital period ($Tp\_callirrhoe$) = **64,601,874.62 seconds**

On the other hand,

Observed orbital period of Jupiter's moon Callirrhoe ($To\_callirrhoe$) gives:

$To\_callirrhoe$ = 64,601,874.62 seconds
(or 758.8 days)

Let's calculate

The percentage margin of error between Jupiter's moon Callirrhoe's Predicted orbital period and its astronomically observed one:

$\Delta[Tp\_callirrhoe / To\_callirrhoe]$ = [64,601,874.62 secs / 65,560,320.00 secs] x 100
= 1.4836%

One can conclude that

The equation (j2) of the orbital period quantizer mechanism predicts correctly Jupiter's moon Callirrhoe's orbital period

Orbital Period Verification Case of Jupiter's Moon Kallichore:

From the astronomically

Observed mean orbital velocity of Jupiter's moon Kallichore ($V_{kallichore}$) as:

$V_{kallichore}$ = 2.28646 km/s or 2,286.46 m
Source:

based on values of:

Kallichore's semi-axis = 24,043,000 km
($\rightarrow$ circumference = km)
Kallichore's sidereal orbital period = 764.7 days
(or 66,070,080 seconds)

Kallichore's semi-axis = 24,043,000 km
($\rightarrow$ circumference = 151,066,624.34 km)
Kallichore's sidereal orbital period = 764.7 days
(or 66,070,080 seconds)

and the equation (s1) we obtain:

Jupiter's moon Kallichore's orbital quantizer value $Dc\_kallichore$ as:

$Dc\_kallichore$ = c / $V_{kallichore}$ = 299,792,458 m / 2,286.46 m $\rightarrow$

$Dc\_kallichore$ = 131,116.42 $\rightarrow$

$Dc\_kallichore^2$ = 17,191,515,593.6164

$Dc\_kallichore^3$ = 2,254,089,979,009,157

That leads to obtain the following Jupiter's moon Kallichore terms of the equation (j2):

[$GM_{jupiter}/c^3$] x $Dc\_kallichore^3$ = 4.70302287823021162e-9 x
  2,254,089,979,009,157
  = 10,601,036.74 seconds

[$GM_{jupiter}/c^3$] x $Dc\_kallichore^3$ x = 10,601,036.74 x 2π
2π = 66,608,278.28 seconds

The outcome of the equation (j2) of the orbital period quantizer mechanism for Jupiter's moon Kallichore yields:

Jupiter's moon Kallichore's Predicted orbital period (Tp_kallichore) = **66,608,278.28 seconds**

On the other hand,

Observed orbital period of Jupiter's moon Kallichore (To_kallichore) gives:

To_kallichore = 66,070,080.00 seconds
(or 764.7 days)

Let's calculate

The percentage margin of error between Jupiter's moon Kallichore's Predicted orbital period and its astronomically observed one:

Δ[Tp_kallichore / To_kallichore] = [66,608,278.28 secs / 66,070,080.00 secs] x 100
= 0.8145%

One can conclude that

The equation (j2) of the orbital period quantizer mechanism predicts correctly Jupiter's moon Kallichore's orbital period

Orbital Period Verification Case of Jupiter's Moon S2003J9:

From the astronomically

Observed mean orbital velocity of Jupiter's moon S2003J9 (Vs2003J9) as:

$V_{s2003J9}$ = 2.29921 km/s or 2,299.21 m
Source:

based on values of:

S2003J9's semi-axis = 24,234,000 km
($\rightarrow$ circumference = km)
S2003J9's sidereal orbital period = 766.5 days
(or 66,225,600 seconds)

and the equation (s1) we obtain:

Jupiter's moon S2003J9's orbital quantizer value Dc_s2003J9 as:

$Dc\_{s2003J9} = c / V_{s2003J9} = 299{,}792{,}458 \text{ m} / 2{,}299.21 \text{ m} \quad \rightarrow$

$Dc\_{s2003J9} = 130{,}389.33 \quad \rightarrow$

$Dc\_{s2003J9}^2 = 17{,}001{,}377{,}377.8489$

$Dc\_{s2003J9}^3 = 2{,}216{,}798{,}205{,}374{,}874$

That leads to obtain the following Jupiter's moon S2003J9 terms of the equation (j2):

$[GM_{jupiter}/c^3] \times Dc\_{s2003J9}^3$ = 4.70302287823021162e-9 × 2,216,798,205,374,874
= 10,425,652.67 seconds

$[GM_{jupiter}/c^3] \times Dc\_{s2003J9}^3 \times 2\pi$ = 10,425,652.67 × 2π
= 65,506,307.67 seconds

The outcome of the equation (j2) of the orbital period quantizer mechanism for Jupiter's moon S2003J9 yields:

Jupiter's moon S2003J9's Predicted orbital period (Tp_s2003J9) = **65,506,307.67 seconds**

On the other hand,

Observed orbital period of Jupiter's moon S2003J9 (To_s2003J9) gives:

To_s2003J9 = 66,225,600.00 seconds (or 766.5 days)

Let's calculate

The percentage margin of error between Jupiter's moon S2003J9's Predicted orbital period and its astronomically observed one:

$\Delta[Tp\_{s2003J9} / To\_{s2003J9}]$ = [65,506,307.67 secs / 66,225,600.00 secs] × 100
= 1.0980%

One can conclude that

The equation (j2) of the orbital period quantizer mechanism predicts correctly Jupiter's moon S2003J9's orbital period

Orbital Period Verification Case of Jupiter's Moon Cyllene:

From the astronomically

Observed mean orbital velocity of Jupiter's moon Cyllene ($V_{cyllene}$)

as:

$V_{cyllene}$ = 2.39998 km/s or 2,399.98 m
Source:

based on values of:

Cyllene's semi-axis = 24,349,000 km
($\rightarrow$ circumference = km)
Cyllene's sidereal orbital period = 737.8 days
(or 63,745,920 seconds)

and the equation (s1) we obtain:

Jupiter's moon Cyllene's orbital quantizer value $D_{c\_cyllene}$ as:

$D_{c\_cyllene}$ = c / $V_{cyllene}$ = 299,792,458 m / 2,399.98 m $\rightarrow$

$D_{c\_cyllene}$ = 124,914.56 $\rightarrow$

$D_{c\_cyllene}^2$ = 15,603,647,299.9936

$D_{c\_cyllene}^3$ = 1,949,122,736,873,888

That leads to obtain the following Jupiter's moon Cyllene terms of the equation (j2):

$[GM_{jupiter}/c^3]$ x $D_{c\_cyllene}^3$ = 4.70302287823021162e-9 x 1,949,122,736,873,888
= 9,166,768.82 seconds

$[GM_{jupiter}/c^3]$ x $D_{c\_cyllene}^3$ x 2π = 9,166,768.82 x 2π
= 57,596,507.16 seconds

The outcome of the equation (j2) of the orbital period quantizer mechanism for Jupiter's moon Cyllene yields:

Jupiter's moon Cyllene's = **57,596,507.16 seconds**
Predicted orbital period
($T_{p\_cyllene}$)

On the other hand,

Observed orbital period of Jupiter's moon Cyllene ($T_{o\_cyllene}$) gives:

$T_{o\_cyllene}$ = 63,745,920.00 seconds
(or 737.8 days)

Let's calculate

The percentage margin of error between Jupiter's moon Cyllene's Predicted orbital period and its astronomically observed one:

Δ[Tp_cyllene / To_cyllene] = [57,596,507.16 secs / 63,745,920.00 secs] x 100
= 10.6767%

One can conclude that

The equation (j2) of the orbital period quantizer mechanism predicts correctly Jupiter's moon Cyllene's orbital period

Evidence Of Orbital Period Quantizer Mechanism Through Moons Of Saturn's Planetary System

Targets Of This Orbital Radius Period Quantizer Mechanism Verification:

Based on my calculations, there appears to exist

A host of evidence of this orbital period quantizer mechanism for a host of known moons that revolve around Saturn.

The planets subjected to this verification are:

Mimas, Enceladus, Tethys, Dione, Rhea, Titan, Hyperion, Iapetus

For each moon therein, we will verify whether or not

The orbital period of each moon of Saturn's planetary system obeys the equation:

$$T = 2\pi \times ([GM/c^3] \times Dc^3) \quad (=j2)$$

The equation of orbital period (j2) shows that time is no other than a mathematical representation of chain of motions, not a pure dimension as Einstein's spacetime paradigm claims.

Reference Data Of This Verification:

From the equations (z1) and (z2) we obtain:

Saturn's effective gravitational energy potential quanta dedicated exclusively to orbital motion of each captive as:

$GM_{saturn} = 3.794072578e+16$ N kg^3 m^2

$GM_{saturn}/c^2 = E_{g1} = 1 \times [GM_{saturn}/c^2]$ $\quad (=z1)$

$GM_{saturn}/c^2 = 0.422147506658319659$ m or 0.000422147506658319659 km

$GM_{saturn}/c^3 = E_{g2} = 1 \times [GM_{saturn}/c^3]$ $\quad (=z3)$

$GM_{saturn}/c^3 = 1.40813251098638264e-9$ m

based on the values of:

Msaturn = 5.6846e+26 kg (as mass of Saturn)
G = 6.6743e-11 (N kg^{-2} m^2)
c = 299,792,458 m

Let's go through each captive subjected to this verification.

Orbital Period Verification Case of Saturn's Moon Mimas:

From the astronomically

Observed mean orbital velocity of Saturn's moon Mimas (Vmimas) as:

Vmimas = 14.31566 km/s or 14,315.66 m
Source:

based on values of:

Mimas' semi-axis = 185,520 km
($\rightarrow$ circumference = km)
Mimas' sidereal orbital period = 0.94242180 days
(or 81,425.24 seconds)

and the equation (s1) we obtain:

Saturn's moon Mimas' orbital quantizer value Dc_mimas as:

Dc_mimas = c / Vmimas = 299,792,458 m / 14,315.66 m $\rightarrow$

Dc_mimas = 20,941.57 $\rightarrow$

Dc_mimas^2 = 438,549,354.0649

Dc_mimas^3 = 9,183,911,996,605

That leads to obtain the following Saturn's moon Mimas terms of the equation (j2):

[GMsaturn/c^3] x Dc_mimas^3 = 1.40813251098638264e-9 x 9,183,911,996,605
= 12,932.16506 seconds

[GMsaturn/c^3] x Dc_mimas^3 x 2π = 12,932.16506 x 2π
= 81,255.18 seconds

The outcome of the equation (j2) of the orbital period quantizer mechanism for Saturn's moon Mimas yields:

Saturn's moon Mimas' = **81,255.18 seconds**
Predicted orbital period

(Tp_mimas)

On the other hand,

Observed orbital period of Saturn's moon Mimas (To_mimas) gives:

To_mimas = 81,425.24 seconds
(or 0.94242180 days)

Let's calculate

The percentage margin of error between Saturn's moon Mimas' Predicted orbital period and its astronomically observed one:

Δ[Tp_mimas / To_mimas] = [81,255.18 secs / 81,425.24 secs] x 100
= 0.20929%

One can conclude that

The equation (j2) of the orbital period quantizer mechanism predicts correctly Saturn's moon Mimas' orbital period

Orbital Period Verification Case of Saturn's Moon Enceladus:

From the astronomically

Observed mean orbital velocity of Saturn's moon Enceladus (Venceladus) as:

Venceladus = 12.63251 km/s or 12,632.51 m
Source:

based on values of:

Enceladus' semi-axis = 238,020 km
($\rightarrow$ circumference = km)
Enceladus' sidereal orbital period = 1.370218 days
(or 118,386.8352 seconds)

and the equation (s1) we obtain:

Saturn's moon Enceladus' orbital quantizer value Dc_enceladus as:

Dc_enceladus = c / Venceladus = 299,792,458 m / 12,632.51 $\rightarrow$ m

Dc_enceladus = 23,731.82 $\rightarrow$

Dc_enceladus^2 = 563,199,280.5124

$Dc\_enceladus^3 = 13,365,743,949,249$

That leads to obtain the following Saturn's moon Enceladus terms of the equation (j2):

$[GMsaturn/c^3] \times Dc\_enceladus^3$ = 1.40813251098638264e-9 × 13,365,743,949,249
= 18,820.73858 seconds

$[GMsaturn/c^3] \times Dc\_enceladus^3 \times 2\pi$ = 18,820.73858 × 2π
= 118,254.18 seconds

The outcome of the equation (j2) of the orbital period quantizer mechanism for Saturn's moon Enceladus yields:

Saturn's moon Enceladus' Predicted orbital period (Tp_enceladus) = **118,254.18 seconds**

On the other hand,

Observed orbital period of Saturn's moon Enceladus (To_enceladus) gives:

To_enceladus = 118,386.8352 seconds
(or 1.370218 days)

Let's calculate

The percentage margin of error between Saturn's moon Enceladus' Predicted orbital period and its astronomically observed one:

$\Delta[Tp\_enceladus / To\_enceladus]$ = [118,254.18 secs / 118,386.8352 secs] × 100
= 0.1121%

One can conclude that

The equation (j2) of the orbital period quantizer mechanism predicts correctly Saturn's moon Enceladus' orbital period

Orbital Period Verification Case of Saturn's Moon Tethys:

From the astronomically

Observed mean orbital velocity of Saturn's moon Tethys (Vtethys) as:

Vtethys = 11.35091 km/s or 11,350.91 m
Source:

based on values of:

Tethys' semi-axis = 294,660 km
($\rightarrow$ circumference = km)
Tethys' sidereal orbital period = 1.887802 days
(or 163,106.0928 seconds)

and the equation (s1) we obtain:

Saturn's moon Tethys' orbital quantizer value Dc_tethys as:

$Dc\_tethys = c / V_{tethys}$ = 299,792,458 m / 11,350.91 m $\rightarrow$

Dc_tethys = 26,411.31 $\rightarrow$

Dc_tethys^2 = 697,557,295.9161

Dc_tethys^3 = 18,423,401,985,201

That leads to obtain the following Saturn's moon Tethys terms of the equation (j2):

$[GM_{saturn}/c^3] \times Dc\_tethys^3$ = 1.40813251098638264e-9 x 18,423,401,985,201
= 25,942.59129 seconds

$[GM_{saturn}/c^3] \times Dc\_tethys^3 \times 2\pi$ = 25,942.59129 x 2π
= 163,002.10 seconds

The outcome of the equation (j2) of the orbital period quantizer mechanism for Saturn's moon Tethys yields:

Saturn's moon Tethys' Predicted orbital period (Tp_tethys) = **163,002.10 seconds**

On the other hand,

Observed orbital period of Saturn's moon Tethys (To_tethys) gives:

To_tethys = 163,106.0928 seconds
(or 1.887802 days)

Let's calculate

The percentage margin of error between Saturn's moon Tethys' Predicted orbital period and its astronomically observed one:

$\Delta[Tp\_tethys / To\_tethys]$ = [163,002.10 secs / 163,106.0928 secs] x 100
= 0.0637%

One can conclude that

The equation (j2) of the orbital period quantizer mechanism predicts correctly Saturn's moon Tethys' orbital period

Orbital Period Verification Case of Saturn's Moon Dione:

From the astronomically

Observed mean orbital velocity of Saturn's moon Dione (Vdione) as:

Vdione = 10.02782 km/s or 10,027.82 m
Source:

based on values of:

Dione's semi-axis = 377,400 km
($\rightarrow$ circumference = km)
Dione's sidereal orbital period = 2.736915 days
(or 236,469.456 seconds)

and the equation (s1) we obtain:

Saturn's moon Dione's orbital quantizer value Dc_dione as:

Dc_dione = c / Vdione = 299,792,458 m / 10,027.82 m $\rightarrow$

Dc_dione = 29,896.07 $\rightarrow$

Dc_dione^2 = 893,775,001.4449

Dc_dione^3 = 26,720,360,007,446

That leads to obtain the following Saturn's moon Dione terms of the equation (j2):

[GMsaturn/c^3] x Dc_dione^3 = 1.40813251098638264e-9 x 26,720,360,007,446
= 37,625.80763 m

[GMsaturn/c^3] x Dc_dione^3 x 2π = 37,625.80763 x 2π
= 236,409.92 seconds

The outcome of the equation (j2) of the orbital period quantizer mechanism for Saturn's moon Dione yields:

Saturn's moon Dione's Predicted orbital period (Tp_dione) = **236,409.92 seconds**

On the other hand,

Observed orbital period of Saturn's moon Dione (To_dione) gives:

To_dione   = 236,469.456 seconds
(or 2.736915 days)

Let's calculate

The percentage margin of error between Saturn's moon Dione's Predicted orbital period and its astronomically observed one:

$\Delta[\text{Tp\_dione} / \text{To\_dione}]$ = [236,409.92 secs / 236,469.456 secs] x 100
= 0.0251%

One can conclude that

The equation (j2) of the orbital period quantizer mechanism predicts correctly Saturn's moon Dione's orbital period

Orbital Period Verification Case of Saturn's Moon Rhea:

From the astronomically

Observed mean orbital velocity of Saturn's moon Rhea (Vrhea) as:

Vrhea = 8.48421 km/s or 8,484.21 m
Source:

based on values of:

Rhea's semi-axis = 527,040 km
($\rightarrow$ circumference = km)
Rhea's sidereal orbital period = 4.517500 days
(or 390,312 seconds)

and the equation (s1) we obtain:

Saturn's moon Rhea's orbital quantizer value Dc_rhea as:

Dc_rhea = c / Vrhea = 299,792,458 m / 8,484.21 m $\rightarrow$

Dc_rhea = 35,335.34 $\rightarrow$

Dc_rhea^2 = 1,248,586,252.9156

Dc_rhea^3 = 44,119,219,766,098

That leads to obtain the following Saturn's moon Rhea terms of the equation (j2):

[GMsaturn/c^3] x Dc_rhea^3 = 1.40813251098638264e-9 x
44,119,219,766,098

$$[GM_{saturn}/c^3] \times Dc\_rhea^3 \times 2\pi \quad \begin{aligned} &= 62{,}125.70771 \text{ seconds} \\ &= 62{,}125.70771 \times 2\pi \\ &= 390{,}347.33 \text{ seconds} \end{aligned}$$

The outcome of the equation (j2) of the orbital period quantizer mechanism for Saturn's moon Rhea yields:

Saturn's moon Rhea's Predicted orbital period (Tp_rhea) $= \mathbf{390{,}347.33}$ **seconds**

On the other hand,

Observed orbital period of Saturn's moon Rhea (To_rhea) gives:

To_rhea $= 390{,}312$ seconds
(or 4.517500 days)

Let's calculate

The percentage margin of error between Saturn's moon Rhea's Predicted orbital period and its astronomically observed one:

$\Delta[Tp\_rhea / To\_rhea] \quad \begin{aligned} &= [390{,}347.33 \text{ secs} / \\ & \quad 390{,}312.00 \text{ secs}] \times 100 \\ &= 0.0090\% \end{aligned}$

One can conclude that

The equation (j2) of the orbital period quantizer mechanism predicts correctly Saturn's moon Rhea's orbital period

Orbital Period Verification Case of Saturn's Moon Titan:

From the astronomically

Observed mean orbital velocity of Saturn's moon Titan (Vtitan) as:

$V_{titan} = 5.57256$ km/s or $5{,}572.56$ m
Source:

based on values of:

Titan's semi-axis = $1{,}221{,}870$ km
($\rightarrow$ circumference = km)
Titan's sidereal orbital period = 15.945421 days
(or $1{,}377{,}684.3744$ seconds)

and the equation (s1) we obtain:

Saturn's moon Titan's orbital quantizer value Dc_titan as:

$Dc\_titan = c / Vtitan = 299{,}792{,}458 \text{ m} / 5{,}572.56 \text{ m}$ →

$Dc\_titan = 53{,}797.97$ →

$Dc\_titan^2 = 2{,}894{,}221{,}576.1209$

$Dc\_titan^3 = 155{,}703{,}245{,}525{,}504$

That leads to obtain the following Saturn's moon Titan terms of the equation (j2):

$[GMsaturn/c^3] \times Dc\_titan^3$ = 1.40813251098638264e-9 x 155,703,245,525,504
= 219,250.80209 seconds

$[GMsaturn/c^3] \times Dc\_titan^3 \times 2\pi$ = 219,250.80209 x 2π
= 1,377,593.41 seconds

The outcome of the equation (j2) of the orbital period quantizer mechanism for Saturn's moon Titan yields:

Saturn's moon Titan's Predicted orbital period (Tp_titan) = **1,377,593.41 seconds**

On the other hand,

Observed orbital period of Saturn's moon Titan (To_titan) gives:

To_titan = 1,377,684.3744 seconds
(or 15.945421 days)

Let's calculate

The percentage margin of error between Saturn's moon Titan's Predicted orbital period and its astronomically observed one:

$\Delta[Tp\_titan / To\_titan]$ = [1,377,593.41 secs / 1,377,684.37 secs] x 100
= 0.0066%

One can conclude that

The equation (j2) of the orbital period quantizer mechanism predicts correctly Saturn's moon Titan's orbital period

Orbital Period Verification Case of Saturn's Moon Hyperion:

From the astronomically

Observed mean orbital velocity of Saturn's moon Hyperion (V_hyperion) as:

V_hyperion = 5.13008 km/s or 5,130.08 m
Source:

based on values of:

Hyperion's semi-axis = 1,500,930 km
(→ circumference = km)
Hyperion's sidereal orbital period = 21.276609 days
(or 1,838,299.0176 seconds)

Hyperion's semi-axis = 1,500,930 km
(→ circumference = 9,430,621.32 km)
Hyperion's sidereal orbital period = 21.276609 days
(or 1,838,299.0176 seconds)

and the equation (s1) we obtain:

Saturn's moon Hyperion's orbital quantizer value $Dc\_hyperion$ as:

$Dc\_hyperion$ = c / V_hyperion = 299,792,458 m / 5,130.08 m →

$Dc\_hyperion$ = 58,438.16 →

$Dc\_hyperion^2$ = 3415018544.1856

$Dc\_hyperion^3$ = 199,567,400,088,085

That leads to obtain the following Saturn's moon Hyperion terms of the equation (j2):

$[GM_{saturn}/c^3] \times Dc\_hyperion^3$	= 1.40813251098638264e-9 x 199,567,400,088,085 = 281,017.34419 seconds
$[GM_{saturn}/c^3] \times Dc\_hyperion^3 \times 2\pi$	= 281,017.34419 x 2π = 1,765,684.04 seconds

The outcome of the equation (j2) of the orbital period quantizer mechanism for Saturn's moon Hyperion yields:

Saturn's moon Hyperion's Predicted orbital period (Tp_hyperion)	= **1,765,684.04 seconds**

On the other hand,

Observed orbital period of Saturn's moon Hyperion (To_hyperion) gives:

$$T_{o\_hyperion} = 1{,}838{,}299.0176 \text{ seconds}$$
$$(\text{or } 21.276609 \text{ days})$$

Let's calculate

The percentage margin of error between Saturn's moon Hyperion's Predicted orbital period and its astronomically observed one:

$$\Delta[T_{p\_hyperion} / T_{o\_hyperion}] = [1{,}765{,}684.04 \text{ secs} / 1{,}838{,}299.01 \text{ secs}] \times 100$$
$$= 4.1125\%$$

One can conclude that

The equation (j2) of the orbital period quantizer mechanism predicts correctly Saturn's moon Hyperion's orbital period

Orbital Period Verification Case of Saturn's Moon Iapetus:

From the astronomically

Observed mean orbital velocity of Saturn's moon Iapetus ($V_{iapetus}$) as:

$V_{iapetus}$ = 3.26423 km/s or 3,264.23 m
Source:

based on values of:

Iapetus's semi-axis = 3,560,850 km
($\rightarrow$ circumference = km)
Iapetus's sidereal orbital period = 79.330183 days
(or 6,854,127.8112 seconds)

and the equation (s1) we obtain:

Saturn's moon Iapetus's orbital quantizer value $D_{c\_iapetus}$ as:

$D_{c\_iapetus} = c / V_{iapetus} = 299{,}792{,}458 \text{ m} / 3{,}264.23 \text{ m}$ $\rightarrow$

$D_{c\_iapetus} = 91{,}841.70$ $\rightarrow$

$D_{c\_iapetus}^2 = 8{,}434{,}897{,}858.89$

$D_{c\_iapetus}^3 = 774{,}675{,}358{,}686{,}817$

That leads to obtain the following Saturn's moon Iapetus terms of the equation (j2):

$[GM_{saturn}/c^3] \times Dc\_iapetus^3$ = 1.40813251098638264e-9 x 774,675,358,686,817
= 1,090,845.55802 seconds

$[GM_{saturn}/c^3] \times Dc\_iapetus^3 \times 2\pi$ = 1,090,845.55802 x 2π
= 6,853,984.78 seconds

The outcome of the equation (j2) of the orbital period quantizer mechanism for Saturn's moon Iapetus yields:

Saturn's moon Iapetus's Predicted orbital period ($Tp\_iapetus$) = **6,853,984.78 seconds**

On the other hand,

Observed orbital period of Saturn's moon Iapetus ($To\_iapetus$) gives:

$To\_iapetus$ = 6,854,127.8112 seconds
(or 21.276609 days)

Let's calculate

The percentage margin of error between Saturn's moon Iapetus's Predicted orbital period and its astronomically observed one:

$\Delta[Tp\_iapetus / To\_iapetus]$ = [6,853,984.78 secs / 6,854,127.81 secs] x 100
= 0.0020%

One can conclude that

The equation (j2) of the orbital period quantizer mechanism predicts correctly Saturn's moon Iapetus's orbital period

Evidence Of Orbital Period Quantizer Mechanism Through Moons Of Uranus' Planetary System

Targets Of This Orbital Radius Period Quantizer Mechanism Verification:

Based on my calculations, there appears to exist

A host of evidence of this orbital period quantizer mechanism for a host of known moons that revolve around Uranus.

The planets subjected to this verification are:

Miranda, Ariel, Umbriel, Titania, Oberon

For each moon therein, we will verify whether or not

The orbital period of each moon of Uranus' planetary system obeys the equation:

$$T = 2\pi \times ([GM/c^3] \times Dc^3) \qquad (=j2)$$

The equation of orbital period (j2) shows that time is no other than a mathematical representation of chain of motions, not a pure dimension as Einstein's spacetime paradigm claims.

Reference Data Of This Orbital Period Quantizer Mechanism Verification:

From the equations (z1) and (z2) we obtain:

Uranus' effective gravitational energy potential quanta dedicated exclusively to orbital motion of each captive as:

$GM_{uranus} = 5.79395983e+15$ N kg^3 m^2

$$GM_{uranus}/c^2 = E_{g1} = 1 \times [GM_{uranus}/c^2] \qquad (=z1)$$

$GM_{uranus}/c^2 = 0.0644664972962191352$ m or
    0.0000644664972962191352 km

$$GM_{uranus}/c^3 = E_{g2} = 1 \times [GM_{uranus}/c^3] \qquad (=z3)$$

$GM_{uranus}/c^3 = 2.15037088412074513e\text{-}10$ m

based on the values of:

Muranus = $8.6810e+25$ kg (as mass of Uranus)
G = 6.6743e-11 (N kg^{-2} m²)
c = 299,792,458 m

Let's go through each captive subjected to this verification.

Orbital Period Verification Case of Uranus' Moon Miranda:

From the astronomically

Observed mean orbital velocity of Uranus' moon Miranda (V$_{miranda}$) as:

V$_{miranda}$ = 6.6832 km/s or 6,683.22 m
Source:

based on values of:

Miranda's semi-axis = 129,900 km
($\rightarrow$ circumference = km)
Miranda's sidereal orbital period = 1.413479 days
(or 122,124.5856 seconds)

and the equation (s1) we obtain:

Uranus' moon Miranda's orbital quantizer value Dc_miranda as:

Dc_miranda = c / V$_{miranda}$ = 299,792,458 m / 6,683.22 m $\rightarrow$

Dc_miranda = 44,857.48 $\rightarrow$

Dc_miranda² = 2,012,193,511.9504

Dc_miranda³ = 90,261,930,218,444

That leads to obtain the following Uranus' moon Miranda terms of the equation (j2):

[GMuranus/c³] x Dc_miranda³ = 2.15037088412074513e-10 x 90,261,930,218,444
= 19,409.66266 m

[GMuranus/c³] x Dc_miranda³ x 2π = 19,409.66266 x 2π
= 121,954.50 seconds

The outcome of the equation (j2) of the orbital period quantizer mechanism for Uranus' moon Miranda yields:

Uranus' moon Miranda's = **121,954.50 seconds**
Predicted orbital period
(Tp_miranda)

On the other hand,

Observed orbital period of Uranus' moon Miranda (To_miranda) gives:

To_miranda = 122,124.5856 seconds
(or 1.413479 days)

Let's calculate

The percentage margin of error between Uranus' moon Miranda's Predicted orbital period and its astronomically observed one:

Δ[Tp_miranda / To_miranda] = [121,954.50 secs / 122,124.58 secs] x 100
= 0.1394%

One can conclude that

The equation (j2) of the orbital period quantizer mechanism predicts correctly Uranus' moon Miranda's orbital period

Orbital Period Verification Case of Uranus' Moon Ariel:

From the astronomically

Observed mean orbital velocity of Uranus' moon Ariel (Variel) as:

Variel = 5.50815 km/s or 5,508.15 m
Source:

based on values of:

Ariel's semi-axis = 190,900 km
($\rightarrow$ circumference = km)
Ariel's sidereal orbital period = 2.520379 days
(or 217,760.7456 seconds)

and the equation (s1) we obtain:

Uranus' moon Ariel's orbital quantizer value Dc_ariel as:

Dc_ariel = c / Variel = 299,792,458 m / 5,508.15 m $\rightarrow$

Dc_ariel = 54,427.06 $\rightarrow$

Dc_ariel^2 = 2,962,304,860.2436

$Dc\_ariel^3 = 161,229,544,366,770$

That leads to obtain the following Uranus' moon Ariel terms of the equation (j2):

$[GMuranus/c^3] \times Dc\_ariel^3$ = 2.15037088412074513e-10 x 161,229,544,366,770
= 34,670.33178 seconds

$[GMuranus/c^3] \times Dc\_ariel^3 \times 2\pi$ = 34,670.33178 x 2π
= 217,840.11 seconds

The outcome of the equation (j2) of the orbital period quantizer mechanism for Uranus' moon Ariel yields:

Uranus' moon Ariel's Predicted orbital period (Tp_ariel) = **217,840.11 seconds**

On the other hand,

Observed orbital period of Uranus' moon Ariel (To_ariel) gives:

To_ariel = 217,760.7456 seconds
(or 2.520379 days)

Let's calculate

The percentage margin of error between Uranus' moon Ariel's Predicted orbital period and its astronomically observed one:

Δ[Tp_ariel / To_ariel] = [217,840.11 secs / 217,760.74 secs] x 100
= 0.0364%

One can conclude that

The equation (j2) of the orbital period quantizer mechanism predicts correctly Uranus' moon Ariel's orbital period

Orbital Period Verification Case of Uranus' Moon Umbriel:

From the astronomically

Observed mean orbital velocity of Uranus' moon Umbriel (Vumbriel) as:

Vumbriel = 4.66777 km/s or 4,667.77 m
Source:

based on values of:

Umbriel's semi-axis = 266,000 km
(→ circumference = km)
Umbriel's sidereal orbital period = 4.144176 days
(or 358,056.8064 seconds)

and the equation (s1) we obtain:

Uranus' moon Umbriel's orbital quantizer value $Dc\_umbriel$ as:

$Dc\_umbriel$ = c / $Vumbriel$ = 299,792,458 m / 4,667.77 m →

$Dc\_umbriel$ = 64,226.05 →

$Dc\_umbriel^2$ = 4,124,985,498.6025

$Dc\_umbriel^3$ = 264,931,524,882,519

That leads to obtain the following Uranus' moon Umbriel terms of the equation (j2):

[GM_{uranus}/c^3] x $Dc\_umbriel^3$ = 2.15037088412074513e-10 x 264,931,524,882,519
= 56,970.10373 seconds

[GM_{uranus}/c^3] x $Dc\_umbriel^3$ x 2π = 56,970.10373 x 2π
= 357,953.71 seconds

The outcome of the equation (j2) of the orbital period quantizer mechanism for Uranus' moon Umbriel yields:

Uranus' moon Umbriel's Predicted orbital period ($Tp\_umbriel$) = **357,953.71 seconds**

On the other hand,

Observed orbital period of Uranus' moon Umbriel ($To\_umbriel$) gives:

$To\_umbriel$ = 358,056.8064 seconds
(or 4.144176 days)

Let's calculate

The percentage margin of error between Uranus' moon Umbriel's Predicted orbital period and its astronomically observed one:

$\Delta[Tp\_umbriel / To\_umbriel]$ = [357,953.71 secs / 358,056.80 secs] x 100
= 0.0287%

One can conclude that

The equation (j2) of the orbital period quantizer mechanism predicts correctly Uranus' moon Umbriel's orbital period

Orbital Period Verification Case of Uranus' Moon Titania:

From the astronomically

Observed mean orbital velocity of Uranus' moon Titania ($V_{titania}$) as:

$V_{titania}$ = 3.64451 km/s or 3,644.51 m
Source:

based on values of:

Titania's semi-axis = 436,300 km
($\rightarrow$ circumference = km)
Titania's sidereal orbital period = 8.705867 days
(or 752,186.9088 seconds)

and the equation (s1) we obtain:

Uranus' moon Titania's orbital quantizer value $Dc\_titania$ as:

$Dc\_titania = c / V_{titania}$ = 299,792,458 m / 3,644.51 m $\rightarrow$

$Dc\_titania$ = 82,258.64 $\rightarrow$

$Dc\_titania^2$ = 6,766,483,854.6496

$Dc\_titania^3$ = 556,601,759,465,433

That leads to obtain the following Uranus' moon Titania terms of the equation (j2):

$[GM_{uranus}/c^3]$ x $Dc\_titania^3$ = 2.15037088412074513e-10 x 556,601,759,465,433
= 119,690.02176 m

$[GM_{uranus}/c^3]$ x $Dc\_titania^3$ x 2π = 119,690.02176 x 2π
= 752,034.58 seconds

The outcome of the equation (j2) of the orbital period quantizer mechanism for Uranus' moon Titania yields:

Uranus' moon Titania's = **752,034.58 seconds**
Predicted orbital period
($Tp\_titania$)

On the other hand,

Observed orbital period of Uranus' moon Titania ($T_{o\_titania}$) gives:

$T_{o\_titania}$ = 752,186.9088 seconds
(or 8.705867 days)

Let's calculate

The percentage margin of error between Uranus' moon Titania's Predicted orbital period and its astronomically observed one:

$\Delta[T_{p\_titania} / T_{o\_titania}]$ = [752,034.58 secs / 752,186.90 secs] x 100
= 0.0202%

One can conclude that

The equation (j2) of the orbital period quantizer mechanism predicts correctly Uranus' moon Titania's orbital period

Orbital Period Verification Case of Uranus' Moon Oberon:

From the astronomically

Observed mean orbital velocity of Uranus' moon Oberon (V_{oberon}) as:

V_{oberon} = 3.15179 km/s or 3,151.79 m
Source:

based on values of:

Oberon's semi-axis = 583,500 km
($\rightarrow$ circumference = km)
Oberon's sidereal orbital period = 13.463234 days
(or 1,163,223.4176 seconds)

and the equation (s1) we obtain:

Uranus' moon Oberon's orbital quantizer value D_{c_oberon} as:

$D_{c\_oberon} = c / V_{oberon}$ = 299,792,458 m / 3,151.79 m $\rightarrow$

D_{c_oberon} = 95,118.15 $\rightarrow$

$D_{c\_oberon}^2$ = 9,047,462,459.4225

$D_{c\_oberon}^3$ = 860,577,891,334,718

That leads to obtain the following Uranus' moon Oberon terms of the equation (j2):

$[GM_{uranus}/c^3] \times D_{c\_oberon}^3$ = 2.15037088412074513e-10 x

$$\frac{[GM_{uranus}/c^3]}{2\pi} \times Dc\_oberon^3 \quad \begin{aligned} x &= \frac{860{,}577{,}891{,}334{,}718}{2\pi} \\ &= 185{,}056.16410 \text{ seconds} \\ x &= 185{,}056.16410 \times 2\pi \\ &= 1{,}162{,}742.17 \text{ seconds} \end{aligned}$$

The outcome of the equation (j2) of the orbital period quantizer mechanism for Uranus' moon Oberon yields:

Uranus' moon Oberon's Predicted orbital period (Tp_oberon) = **1,162,742.17 seconds**

On the other hand,

Observed orbital period of Uranus' moon Oberon (To_oberon) gives:

To_oberon = 1,163,223.4176 seconds
(or 13.463234 days)

Let's calculate

The percentage margin of error between Uranus' moon Oberon's Predicted orbital period and its astronomically observed one:

Δ[Tp_oberon / To_oberon] = [1,162,742.17 secs / 1,163,223.41 secs] x 100
= 0.0413%

One can conclude that

The equation (j2) of the orbital period quantizer mechanism predicts correctly Uranus' moon Oberon's orbital period

Examination Of Stars Of Milky Way Rotating Disk With Potential Gravitational Energy Level Jump In Orbital Period Creation's Quantizer Mechanism

Verification Cases of Milky Way Stars That May Reveal Orbital Period Creation's Quantizer Mechanism

The correlations between the orbital velocity of stars and their orbital radii inside the Milky Way are following diagram:

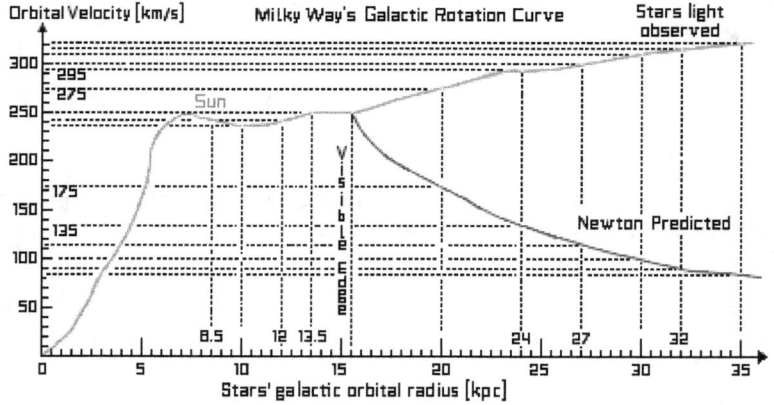

The verification here makes use of the

The discovered equations of orbital period quantizer mechanism:

$$T = 2\pi \times ([GM/c^3] \times Dc^3) \qquad (=j2)$$

which is same as

$$T / Dc^3 = [GM/c^3] \times 2\pi \qquad (=j3)$$

to deduce an equation to reveal the acting mass for a given orbital velocity and orbital radius of a captive:

$$M = [T \times c^3] / [Dc^3 \times 2\pi \times G] \qquad (=j5)$$

As the mass of the Sun is known as:

$M_{sun} = 1.9885e+30$ kg

We can deduce:

The Sun's effective gravitational energy potential quanta dedicated exclusively to orbital motion of each captive (E_{g1}) as:

$GM_{sun}/c^3 = 4.92571420697396808e-6$ second

    based on the values of:

    $M_{sun} = 1.9885e+30$ kg (as mass of the Sun)
    $G = 6.6743e-11$ (N kg^{-2} m^2)
    $c = 299,792,458$ m

All the studies in this segment will use the following public records with respect to the observed astronomical data about the Milky Way:

Visible Mass Only (First estimate method):

    Milky Way's Total visible mass: 100 Billion suns
      (estimated 250 billion stars with
       average star's mass = 0.4 solar mass)
    hence its GM value:

    $GM_{mw100Bsuns}/c^3$ = $[GM_{sun}/c^3 \times 1.e+11]$
                             = $[4.92571420697396808e-6 \times 1.e+11]$

                               = 492,571.4206973968 seconds

Visible Mass Only (Second estimate method):

    Milky Way's Total visible mass: 91 Billion suns
      (estimated by equation $M = a^3 / p^2$
       with a = 8 kpc and p = 220 million years)
    hence its GM value::

    $GM_{mw91Bsuns}/c^3$ = $[GM_{sun}/c^3 \times 9.1e+10]$
                             = $[4.92571420697396808e-6 \times 1.e+11]$

                             = 448,239.99283463109528 secs

Visible Mass Only (Third estimate method):

    Milky Way's Total visible mass: 200 Billion suns
      (estimated value 2022)

hence its GM value:

$GM_{mw200Bsuns}/c^3$ = $[GM_{sun}/c^3 \times 2.e+11]$
= $[4.92571420697396808e-6 \times 2.e+11]$

= 985,142.841394793616 secs

Milky Way Center Bulge's visible mass: 10 Billion suns
(estimated value 2022)
hence its GM value:

$GM_{mwctr10Bsuns}/c^3$ = $[GM_{sun}/c^3 \times 1.e+10]$
= $[4.92571420697396808e-6 \times 1.e+10]$

= 49,257.1420697396808 secs

Milky Way's Rotating Disk's visible mass: 190 Billion suns
(first estimated value 2022)
hence its GM value:

$GM_{mwrd190Bsuns}/c^3$ = $[GM_{sun}/c^3 \times 1.9e+11]$
= $[4.92571420697396808e-6 \times 1.e+10]$

= 935,885.6993250539352 secs

Milky Way's Rotating Disk's visible mass: 150 Billion suns
(second estimated value 2022)
hence its GM value:

$GM_{mwrd150Bsuns}/c^3$ = $[GM_{sun}/c^3 \times 1.5e+11]$
= $[4.92571420697396808e-6 \times 1.e+10]$

= 738,857.131046095212 secs

Visible mass and Dark matter:

Milky Way's Total Visible and Dark matter   1.54 trillion suns
mass estimate #1

Other estimated value of 2019 is   1.29 trillion suns

Milky Way's Total Estimated Visible and   1.54 trillion suns
Dark matter mass is
Milky Way's Total Visible and Dark matter
mass estimate #2
Other estimated value of 2019 is   1.29 trillion suns

Orbital Period Verification Case of Milky Way Stars At 13.5 Kpc And 250 Km Velocity:

Let's examine the case of a hypothetical high orbital velocity of 250 km/second for a hypothetical star moving inside the rotating disk of the Milky Way at an orbital radius of 13.5 kpc.

$V_{mwstar13.5kpc}$ = 250 km/s or 250,000 m/s

$R_{mwstar13.5kpc}$ = 13.5 kpc
= 4.1715e+17 km
(=13500 parsecs x 3.09e+13 km)

Because we already have the observed orbital radius and velocity of the typical stars then we can deduce the required mass that causes the observed data.

The orbital period of the Milky Way stars at 13.5 kpc (denoted as $T_{mwstar13.5kpc}$) from the observed data is:

$T_{o\_mwstar13.5kpc}$ = [$R_{mwstar13.5kpc}$ x 2π] / $V_{mwstar13.5kpc}$
= 4.1715e+20 m x 2π] / 250,000 [m/s]
= 10,484,123,003,559,858 seconds
  or 332,449,359.5750 years

The orbital quantizer of the said stars at 13.5 kpc from the observed data is:

$D_{cmwstar13.5kpc}$ = 299,792,458 m / 250,000 m
= 1,199.1698

$D_{cmwstar13.5kpc}^3$ = $1,199.1698^3$
= 1,724,416,016

We can calculate the predicted orbital period of all said stars at 13.5 kpc via the equation (j2) as:

$T_{p\_mwstar13.5kpc}$ =   (=j2)
   2π x ([$GM_{mw100Bsuns}/c^3$] x $D_{cmwstar13.5kpc}^3$)

That leads to obtain the following all Milky Way stars at 13.5 kpc terms of the equation (j2):

[$GM_{mw100Bsuns}/c^3$] x $D_{cmwstar13.5kpc}^3$
= 492,571.4206973968 secs x 1,724,416,016
= 849,398,046,874,464 secs

[$GM_{mw100Bsuns}/c^3$] x = 849,398,046,874,464 x 2π

$D_{cmwstar13.5kpc}{}^3 \times 2\pi$ = 5,336,925,328,068,669 secs

The outcome of the equation (j2) of the orbital period quantizer mechanism for Milky Way stars at 13.5 kpc yields:

Milky Way stars at 13.5 kpc's Predicted orbital period ($T_{p\_mwstar13.5kpc}$) = **5,336,925,328,068,669 secs**

On the other hand,

Observed orbital period of Milky Way stars at 13.5 kpc ($T_{o\_mwstar13.5kpc}$) gives:

$T_{o\_mwstar13.5kpc}$ = 10,484,123,003,559,858 seconds (or 332,449,359 years)

Let's calculate

The percentage margin of error between the Predicted orbital period of Milky Way stars at 13.5 kpc's and its astronomically observed one:

$\Delta[T_{p\_mwstar13.5kpc} / T_{o\_mwstar13.5kpc}]$ = [5,336,925,328,068,669 / 10,484,123,003,559,858] x 100
= 196.4450%

One can conclude that

The equation (j2) of the orbital period quantizer mechanism does not predict correctly the 's orbital period of Milky Way stars at 13.5 kpc

Let's derive the

Acting gravitational mass for Milky Way stars at 13.5 kpc from the equation (j3)

By

Plugging the obtained values of Milky Way stars at 13.5 kpc into the equation (j3), we get:

$T_{mwstar13.5kpc} / D_{cmwstar13.5kpc}{}^3 = [GM_{mw13.5kpc}/c^3] \times 2\pi$

The expected mass of the Milky Way at the radius of 13.5 kpc from the equation (j5) is:

$M_{mw13.5kpc}$ = $[T_{mwstar13.5kpc} \times c^3]$ / $[D_{cmwstar13.5kpc}{}^3 \times 2\pi \times G]$

= [10,484,123,003,559,858 x 26,944,002,417,373,989,539,335,912] /

$$[1{,}724{,}416{,}016 \times 2\pi \times 6.6743\text{e-}11]$$

$$= 2.8248423555196306588861917537877\text{e+}41 / 0.72314875001673056931925737510191$$
$$= 3.90630884096014\text{e+}41 \text{ kg}$$

$M_{mw13.5kpc}$ = 196,445,000,802 Suns
(=3.90630884096014e+41 kg / 1.9885e+30 kg)

Let's calculate

The percentage ratio between the predicted acting mass and the observed one of Milky Way stars at 13.5 kpc:

$\Delta[M_{mw13.5kpc} / M_{mw100Bsuns}]$ = [196,445,000,802 suns / 100,000,000,000 suns] x 100
= 196.4450%

One can conclude that

The ratios between the predicted mass and the observed mass and

The ratios between the predicted orbital period and the observed orbital period regarding

Milky Way stars at 13.5 kpc is basically the same

Orbital Period Verification Case of Milky Way Stars At 15.5 Kpc And 250 Km Velocity:

Let's examine the case of a hypothetical high orbital velocity of 250 km/second for a hypothetical star moving inside the rotating disk of the Milky Way at an orbital radius of 15.5 kpc.

$V_{mwstar15.5kpc}$ = 250 km/s or 250,000 m/s

$R_{mwstar15.5kpc}$ = 15.5 kpc
= 4.7895e+17 km
(=15500 parsecs x 3.09e+13 km)

Because we already have the observed orbital radius and velocity of the typical stars then we can deduce the required mass that causes the observed data.

The orbital period of the Milky Way stars at 15.5 kpc (denoted as $T_{o\_mwstar15.5kpc}$) from the observed data is:

$T_{o\_mwstar15.5kpc}$ = $[R_{mwstar15.5kpc} \times 2\pi] / V_{mwstar15.5kpc}$
= 4.7895e+20 m x 2π] / 250,000 [m/s]

= 12,037,326,411,494,651 seconds
or 381,701,116.54917 years

The orbital quantizer of the said stars at 15.5 kpc from the observed data is:

Dcmwstar15.5kpc = 299,792,458 m / 250,000 m
= 1,199.1698

Dcmwstar15.5kpc^3 = 1,199.1698^3
= 1,724,416,016

We can calculate the predicted orbital period of all said stars at 15.5 kpc via the equation (j2) as:

Tp_mwstar15.5kpc = (=j2)
2π x ([GMmw100Bsuns/c^3] x Dcmwstar15.5kpc^3)

That leads to obtain the following all Milky Way stars at 15.5 kpc terms of the equation (j2):

[GMmw100Bsuns/c^3] x Dcmwstar15.5kpc^3
= 492,571.4206973968 secs x 1,724,416,016
= 849,398,046,874,464 secs

[GMmw100Bsuns/c^3] x Dcmwstar15.5kpc^3 x 2π
= 849,398,046,874,464 x 2π
= 5,336,925,328,068,669 secs

The outcome of the equation (j2) of the orbital period quantizer mechanism for Milky Way stars at 15.5 kpc yields:

Milky Way stars at 15.5 kpc's Predicted orbital period (Tp_mwstar15.5kpc) = **5,336,925,328,068,669 secs**

On the other hand,

Observed orbital period of Milky Way stars at 15.5 kpc (To_mwstar15.5kpc) gives:

To_mwstar15.5kpc = 12,037,326,411,494,651 seconds
(or 381,701,116.54917 years)

Let's calculate

The percentage margin of error between the Predicted orbital period of Milky Way stars at 15.5 kpc's and its astronomically observed one:

Δ[Tp_mwstar15.5kpc / = [5,336,925,328,068,669 /

To_mwstar15.5kpc] 12,037,326,411,494,651] x 100
= 225.5479%

One can conclude that

The equation (j2) of the orbital period quantizer mechanism does not predict correctly the 's orbital period of Milky Way stars at 15.5 kpc

Let's derive the

Acting gravitational mass for Milky Way stars at 15.5 kpc from the equation (j3)

By

Plugging the obtained values of Milky Way stars at 15.5 kpc into the equation (j3), we get:

$T_{mwstar15.5kpc} / D_{cmwstar15.5kpc}^3 = [GM_{mw15.5kpc}/c^3] \times 2\pi$

The expected mass of the Milky Way at the radius of 15.5 kpc from the equation (j5) is:

$M_{mw15.5kpc}$ = $[T_{mwstar15.5kpc} \times c^3]$ / $[D_{cmwstar15.5kpc}^3 \times 2\pi \times G]$

= [12,037,326,411,494,651 x 26,944,002,417,373,989,539,335,912] / [1,724,416,016 x 2π x 6.6743e-11]

= 3.2433375193003164728604235821799e+41 / 0.7231487500167305693192573751019

= 4.485021261843e+41 kg

$M_{mw15.5kpc}$ = 225,547,963,884 Suns
(= 4.485021261843e+41 / 1.9885e+30 kg)

Let's calculate

The percentage ratio between the predicted acting mass and the observed one of Milky Way stars at 15.5 kpc:

$\Delta[M_{mw15.5kpc} / M_{mw100Bsuns}]$ = [225,547,963,884 suns / 100,000,000,000 suns] x 100
= 225.5479%

One can conclude that

The ratios between the predicted mass and the observed mass and

The ratios between the predicted orbital period and the observed orbital period regarding

Milky Way stars at 15.5 kpc is basically the same

Orbital Period Verification Case of Milky Way Stars At 20 Kpc And 275 Km Velocity:

Let's examine the case of a hypothetical high orbital velocity of 275 km/second for a hypothetical star moving inside the rotating disk of the Milky Way at an orbital radius of 20 kpc.

$V_{mwstar20kpc}$ = 275 km/s or 275,000 m/s

$R_{mwstar20kpc}$ = 20 kpc
= 6.18e+17 km
(=20000 parsecs x 3.09e+13 km)

Because we already have the observed orbital radius and velocity of the typical stars then we can deduce the required mass that causes the observed data.

The orbital period of the Milky Way stars at 20 kpc (denoted as To_mwstar20kpc) from the observed data is:

To_mwstar20kpc = [$R_{mwstar20kpc}$ x 2π] / $V_{mwstar20kpc}$
= 6.18e+20 m x 2π] / 275,000 [m/s]
= 14,120,030,981,225,397 seconds
or 447,743,245.2189 years

The orbital quantizer of the said stars at 20 kpc from the observed data is:

$Dc_{mwstar20kpc}$ = 299,792,458 m / 275,000 m
= 1090.15439

$Dc_{mwstar20kpc}^3$ = 1090.15439^3
= 1,295,579,370

We can calculate the predicted orbital period of all said stars at 20 kpc via the equation (j2) as:

Tp_mwstar20kpc = (=j2)
2π x ([$GM_{mw100Bsuns}/c^3$] x $Dc_{mwstar20kpc}^3$)

That leads to obtain the following all Milky Way stars at 20 kpc terms of the equation (j2):

[GMmw100Bsuns/c^3] x Dcmwstar20kpc^3
= 492,571.4206973968 secs x 1,295,579,370
= 638,165,370,907,138 secs

[GMmw100Bsuns/c^3] x Dcmwstar20kpc^3 x 2π
= 638,165,370,907,138 x 2π
= 4,009,711,282,034,540 secs

The outcome of the equation (j2) of the orbital period quantizer mechanism for Milky Way stars at 20 kpc yields:

Milky Way stars at 20 kpc's Predicted orbital period (Tp_mwstar20kpc) = **4,009,711,282,034,540 secs**

On the other hand,

Observed orbital period of Milky Way stars at 20 kpc (To_mwstar20kpc) gives:

To_mwstar20kpc = 14,120,030,981,225,397 seconds (or 447,743,245.2189 years)

Let's calculate

The percentage margin of error between the Predicted orbital period of Milky Way stars at 20 kpc's and its astronomically observed one:

Δ[Tp_mwstar20kpc / To_mwstar20kpc]
= [4,009,711,282,034,540 / 14,120,030,981,225,397] x 100
= 352.145%

One can conclude that

The equation (j2) of the orbital period quantizer mechanism does not predict correctly the 's orbital period of Milky Way stars at 20 kpc

Let's derive the

Acting gravitational mass for Milky Way stars at 20 kpc from the equation (j3)

By

Plugging the obtained values of Milky Way stars at 20 kpc into the equation (j3), we get:

To_mwstar20kpc / Dcmwstar20kpc^3 = [GMmw20kpc/c^3] x 2π

The expected acting mass of Milky Way stars at the radius of 20 kpc from the equation (j5) is:

$M_{mw20kpc}$ = [$T_{mwstar20kpc}$ x c^3] /
[$D_{cmwstar20kpc}^3$ x 2π x G]

= [14,120,030,981,225,397 x
26,944,002,417,373,989,539,335,912] /
[1,295,579,370 x 2π x 6.6743e-11]

= 3.80450148891532722271861840726094e+41 /
0.54331239867292167417353933860845
= 7.00241978318199e+41 kg

$M_{mw20kpc}$ = 352,145,827,668 Suns
(= 7.00241978318199e+41 / 1.9885e+30 kg)

Let's calculate

The percentage ratio between the predicted acting mass and the observed one of Milky Way stars at 20 kpc:

Δ[$M_{mw20kpc}$ / $M_{mw100Bsuns}$] = [352,145,827,668 suns /
100,000,000,000 suns] x 100
= 352.1458%

One can conclude that

The ratios between the predicted mass and the observed mass and

The ratios between the predicted orbital period and the observed orbital period regarding

Milky Way stars at 20 kpc is basically the same

But in my discovered orbital quantizer mechanism,

There is a big difference between the origins of the predicted orbital period and the predicted mass:

The orbital periods from one orbit to the next can increase proportionally while maintaining the same orbital velocity as long as any new circumference generated by a new orbital period maintains the 2π ratio from one orbital period to the next.

Orbital Period Verification Case of Milky Way Stars At 27 Kpc And 300 Km Velocity:

Let's examine the case of a hypothetical high orbital velocity of 300 km/second for a hypothetical star moving inside the rotating disk of the Milky Way at an orbital radius of 27 kpc.

$V_{mwstar27kpc}$ = 300 km/s or 300,000 m/s

$R_{mwstar27kpc}$ = 27 kpc
= 8.343e+17 km
(=30000 parsecs x 3.09e+13 km)

Because we already have the observed orbital radius and velocity of the typical stars then we can deduce the required mass that causes the observed data.

The orbital period of the Milky Way stars at 27 kpc (denoted as $T_{o\_wstar27kpc}$) from the observed data is:

$T_{o\_wstar27kpc}$ = [$R_{mwstar27kpc}$ x 2π] / $V_{mwstar27kpc}$
= 8.343e+20 m x 2π] / 300,000 [m/s]
= 17,473,538,339,266,430 seconds
or 554,082,265.9584 years

The orbital quantizer of the said stars at 27 kpc from the observed data is:

$D_{cmwstar27kpc}$ = 299,792,458 m / 300,000 m
= 999.3081933

$D_{cmwstar27kpc}^3$ = 999.3081933^3
= 997,926,015

We can calculate the predicted orbital period of all said stars at 27 kpc via the equation (j2) as:

$T_{p\_mwstar27kpc}$ = (=j2)
2π x ([$GM_{mw100Bsuns}/c^3$] x $D_{cmwstar27kpc}^3$)

That leads to obtain the following all Milky Way stars at 27 kpc terms of the equation (j2):

[$GM_{mw100Bsuns}/c^3$] x $D_{cmwstar27kpc}^3$ = 492,571.4206973968 x 997,926,015
= 491,549,834,959,441 secs

[$GM_{mw100Bsuns}/c^3$] x $D_{cmwstar27kpc}^3$ x 2π = 491,549,834,959,441 x 2π
= 3,088,498,700,763,710 secs

The outcome of the equation (j2) of the orbital period quantizer mechanism for Milky Way stars at 27 kpc yields:

Milky Way stars at 27 kpc's = **3,088,498,700,763,710 secs**

Predicted orbital period
(Tp_mwstar27kpc)

On the other hand,

Observed orbital period of Milky Way stars at 27 kpc (To_mwstar27kpc) gives:

To_mwstar27kpc = 17,473,538,339,266,430 seconds
(or 447,743,245.2189 years)

Let's calculate

The percentage margin of error between the Predicted orbital period of Milky Way stars at 27 kpc's and its astronomically observed one:

Δ[Tp_mwstar27kpc / To_mwstar27kpc] = [3,088,498,700,763,710 secs / 17,473,538,339,266,430 secs] x 100
= 565.76%

One can conclude that

The equation (j2) of the orbital period quantizer mechanism does not predict correctly the 's orbital period of Milky Way stars at 27 kpc

Let's derive the

Acting gravitational mass for Milky Way stars at 27 kpc from the equation (j3)

By

Plugging the obtained values of Milky Way stars at 27 kpc into the equation (j3), we get:

Tmwstar27kpc / Dcmwstar27kpc^3 = [GMmw27kpc/c^3] x 2π

The expected mass of the Milky Way at the radius of 27 kpc from the equation (j5) is:

Mmw27kpc = [Tmwstar27kpc x c^3] / [Dcmwstar27kpc^3 x 2π x G]

= [17,473,538,339,266,430 x 26,944,002,417,373,989,539,335,912] / [997,926,015 x 2π x 6.6743e-11]

= 4.70807059253271776481031958964545e+41 / 0.418488893434417695856895691865811
= 1.12501685621688e+42

$M_{mw27kpc}$ = 565,761,557,061 Suns
(= 1.12501685621688e+42 / 1.9885e+30 kg)

According to Newton's law of orbital velocity of celestial objects, we must have:

$M_{mw27kpc\_n}$ = $R_{mw27kpc}$ x $V_{mw27kpc}^2$ / G
= [8.343e+20 x 300,000²] / 6.6743e-11
= 1.12501685570022e+42 kg

Hence

$M_{mw27kpc\_n}$ = 565,761,556,801 Suns
(= 1.12501685570022e+42 / 1.9885e+30 kg)

Let's calculate

The percentage ratio between the predicted acting mass and the observed one of Milky Way stars at 27 kpc:

$\Delta[M_{mw27kpc} / M_{mw100Bsuns}]$ = [565,761,556,801 suns / 100,000,000,000 suns] x 100
= 565.7615%

One can conclude that

The ratios between the predicted mass and the observed mass and

The ratios between the predicted orbital period and the observed orbital period regarding

Milky Way stars at 27 kpc is basically the same

Revelation Of Deep Quantizer Mechanism Of Orbital Period Creation For Galactic Rotating Disk Hosted Stars

Gradual Change Of Deduced Milky Way's Mass Values Between Galactic Radii 13.5 Kpc And 15.5 Kpc:

Let's calculate subsequently

The deduced mass change of the Milky Way for every parsec advancement along a radius segment of 2 kilo-parsecs (from the galaxy center out) between 13.5 kpc and 15.5 kpc, based on the equation (j5):

$M_{mw15.5kpc} - M_{mw13.5kpc}$ = 5.7871242088286e+40 kg

(=4.485021261843e+41 kg - 3.90630884096014e+41 kg)

The deduced mass change of the Milky Way for every parsec advancement along a radius segment of 2 kilo-parsecs (from the galaxy center out) between 13.5 kpc and 15.5 kpc (denoted as $\Delta M_{mw15.5-13.5kpc\_perparsec}$):

$\Delta M_{mw15.5-13.5kpc\_perparsec}$ = 2.8935621044143e+37 kg

(= 5.7871242088286e+40 kg / 2000)

The mass change ratio of the Milky Way for every parsec advancement along a radius segment of 2 kilo-parsecs (from the galaxy center out) between 13.5 kpc and 15.5 kpc (denoted as Ratio$\Delta M_{mw15.5-13.5kpc\_perparsec}$) is:

Ratio$\Delta M_{mw15.5-13.5kpc\_perparsec}$ = 13,500.0000000028857

(= $M_{mw13.5kpc}$ / $\Delta M_{mw15.5-13.5kpc}$
= 3.90630884096014e+41 kg / 2.8935621044143e+37 kg)

There is a relevant fact that

The equation (j5)

$$M = [T \times c^3] / [Dc^3 \times 2\pi \times G] \qquad (=j5)$$

shows that the required acting mass for a specific orbital period of a captive are exclusively interdependent.

This is because all other values inside the right-handed side of the equation (j5) $2\pi \times$, G and c^3 are constant values, and there are no other variables inside the equation either.

The orbital quantizer Dc inside the right-handed side of the equation (j5) does not change either because it depends solely on the constant orbital velocity of the stars inside the rotating disk-owned torus segment.

By the same token for the following specific case,

The mass change ratio of the Milky Way for every parsec advancement along a radius segment of 2 kilo-parsecs (from the galaxy center out) between 13.5 kpc and 15.5 kpc is proportionally identical to the orbital period change ratio therefrom, as shown by the following equation:

$$M_{mwkpc} = [T_{mwstarkpc} \times c^3] / [D_{cmwstarkpc}^3 \times 2\pi \times G]$$

where $D_{cmwstarkpc}$ remains constant

Then one must conclude that

The orbital period change ratio of the Milky Way stars for every parsec advancement along the typical radius from the galaxy center out between 13.5 kpc and 15.5 kpc can be solely defined by the following shortened equation:

$$Ratio\Delta M_{mw}15.5\text{-}13.5kpc = Ratio\Delta[T_{mwstarkpc} \times c^3]$$

which is same as

$$Ratio\Delta M_{mw}15.5\text{-}13.5kpc = Ratio\Delta[T_{mwstarkpc}]$$

Based on my analysis,

This ratio equivalence finding leads to the revelation of the following galactic features of the Milky Way:

The galactic rotating disk must be made of organized galactic concentric orbits.

Each said organized galactic concentric orbit hosts a set of stars that share the same orbital quantizer (hence the same velocity) **and roughly the same equal distance from one star to the next** (hence no collisions among them while revolving around the galaxy center)**.**

The galactic radial distance progress inside a rotating disk-owned torus segment (from the first organized galactic concentric orbit to the last thereof) is defined by an incremental progress of a specific equal discrete values that must maintain the same orbital quantizer.

The requirement of specific equal discrete values of this galactic radial distance progress that maintains the same orbital quantizer is a fundamental element of this galactic quantum mechanism. This is because, without it, any random value thereof can generate a random orbital period value hence a random orbital velocity (then a random orbital quantizer) which in turn breaks the observed galactic flat rotation curve phenomenon.

The expected continuing mass increase of the galaxy that originates from Newton's law equation "$V^2=GM/R$" appears to be irrelevant to the galactic radial distance progress of this galactic orbital quantizer mechanism. This is because the galactic radial distance progress can solely depend on the quantized progress of the orbital periods.

This is because the equation

$$T = 2\pi \times ([GM/c^3] \times Dc^3) \qquad (=j2)$$

Can be rewritten as

$$T = 2\pi \times n \times ([GM/c^3] \times Dc^3) \qquad (j7)$$

where "n" is an integer of values from 1 to infinite.

In this new equation, the value of "n" can be multiplied with any multiplicand "$GM/c3$", "$Dc3$" or 2π. In this mathematical constraint, the value of "n" does not depend solely on the gravitational mass (GM) as Newton's law equations "$V^2=GM/R$" and "$F=GMm/R^2$" require.

By the same token and consequently, it means that, in each torus segment of the galactic rotating disk, the galaxy can just maintain the same gravitational energy (GM/c3) while incrementing the multiplier "n" for the cube value of the orbital quantizer (Dc3). In such setting, each subsequent galactic orbit (inside a given torus segment of the galactic rotating disk) can force its attached stars to maintain the same orbital quantizer, hence the same velocity.

For instance, a galactic orbital period obtained with "n=2" is the same as two rounds of the reference orbital period obtained with "n=1".

This galactic orbital quantizer mechanism appears to be one piece of evidence that dark matter is not needed to maintain the same orbital velocity of all stars inside a torus segment of the galactic rotating disk, so long that the orbital periods of the involved stars obey this mechanism.

List of Observed Orbital Velocities Vs. Orbital Radii of Stars Inside Milky Way:

The public known astronomical data that were used for all the galactic flat rotation curve verification cases of the Milky Way in this chapter are as follows:

Milky Way Galactic Radius	Stars' Observed Orbital Velocity
8.5 kpc	220 km/s

10 kpc	210 km/s ?
12 kpc	220 km/s ?
13.5 kpc	250 km/s ?
15.5 kpc	250 km/s ? (Visible edge)
20 kpc	275 km/s
24 kpc	295 km/s
27 kpc	300 km/s
30 kpc	310 km/s ?
32 kpc	315 km/s ?
35 kpc	320 km/s

Previous Discovery Claim Of Orbital Velocity-Radius Quantizer Mechanism

My previous discovery claim hereby pertains to an

Orbital velocity-radius quantizer mechanism

A comprehensive insight into the components of this orbital quantizer mechanism and their relationship can be visualized via the following diagram:

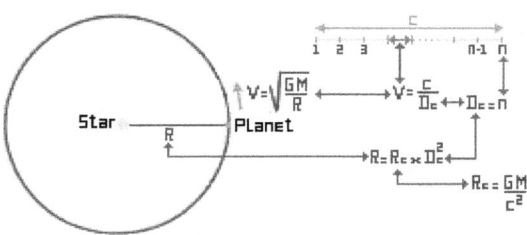

Orbital Velocity-Radius Quantizer Mechanism's Nature

My previous discovery claim hereby postulates that:

The fundamental law that determines the values of orbital velocity and orbital radius of any captive orbiting around a captor is a quantum law of gravitation characterized by an orbital velocity-radius quantizer mechanism.

And by the same token,

The classical emergent law of this orbital velocity-radius quantizer mechanism is no other than Newton's law of orbital velocity of captives, known via the equation:

$$V^2 = GM/R \tag{u1}$$

where:
V is the actual orbital velocity of a specific captive around its captor.
G is Newton's gravitational constant
M is the mass of the captor

Orbital Velocity-Radius Quantizer Mechanism's Components

This orbital velocity-radius quantizer mechanism comprises the two quantum components and their equations as follows:

Quantum Core Radius Component:

The first component of this orbital velocity-radius quantizer mechanism is:

Quantum Core Radius (Rc for short):

Quantum core radius is a gravitational energy quanta generated by the gravitational mass of a captor in order to determine the final radius values of all of its captives.

Quantum core radius' equation is:

$$\mathbf{R_c = GM/c^2} \qquad (w2)$$

where:

G is Newton's gravitational constant
M is the mass of the captor
c is the speed of light in vacuum

Quantum core radius' value must be the same as the said effective orbital motion energy potential that every captive receives from its captor, hence:

$$\mathbf{R_c = E_{g1}} \qquad (w1)$$

Quantum core radius of a captive should not be confused with its classical orbital radius.

The quantum core radius is the common base of the calculation of the actual orbital radius of each captive of the same captor.

Orbital Quantizer Component:

The second component of this orbital velocity-radius-period-precession quantizer mechanism is:

Orbital Quantizer (Dc for short):

Orbital quantizer is for each captive alone.

Orbital quantizer is created by a captor uniquely for a specific captive to determine the latter's final orbital radius and orbital velocity.

Dc is specified by the two equivalent equations.

Orbital Quantizer's First Equation:

$$Dc = c / V \qquad (s1)$$

where:
c is the speed of light in vacuum

V is the actual orbital velocity of a specific captive around its captor.

Relationship between c, V and Dc can be visualized by the following diagram:

$$V = \frac{c}{Dc} \leftrightarrow Dc = n$$

Orbital Quantizer's Second Equation:

$$Dc = \sqrt{[R / Rc]} \qquad (s2)$$

where:
R is the actual orbital radius of a specific captive around its captor.
Rc is the quantum core radius, previously defined by (w2)

And for low cosmic systems, (s2) is same as:

$$Dc = \sqrt{(R / [GM/c^2])} \qquad (s3)$$

Dc is the bridge between the orbital velocity and the orbital radius of the same captive

Orbital Velocity-Radius Quantizer Mechanism's Equations

This orbital velocity-radius quantizer mechanism postulates the two following equations that create the functional links between the orbital velocity and the orbital radius of a captive with respect to its captor.

Equation Linking Orbital Velocity Of A Captive To Its Orbital Quantizer:

Orbital Velocity Quantizer Equation:

Based on my previous discovered equation:

$$R_c = GM/c^2 \qquad (=w2)$$

We can deduce:

$$c^2 = GM / R_c \qquad (w2.1)$$

And the speed of light "c" is just a particular case of Newton's law of orbital velocity of captives "$V^2 = GM/R$".

Based on my discovered equation of orbital quantizer D_c:

$$D_c = c / V \qquad (=s1)$$

The equation (w2.1) can be rewritten as:

$$c^2 = (V \times [c/V])^2 = GM / R_c \qquad (w2.2)$$

Hence

$$\mathbf{(V \times D_c)^2 = GM / R_c} \qquad (x1)$$

> where:
> V is the actual orbital velocity of the captive
>
> D_c is the orbital quantizer generated by the captor for this captive
>
> G is Newton's gravitational constant
>
> M is the mass of the captor
>
> R_c is the quantum core radius, previously defined by (w2)

Inside the equation (x1), we can move the D_c term to the other side of the equation, hence:

$$\mathbf{V^2 = GM / [R_c \times D_c^2]} \qquad (v1)$$

And this equation (v1) is the quantum version of Newton-discovered law of classical orbital velocity of captives:

$$V^2 = GM/R \qquad (=u1)$$

The equation (x1) and its derived one (v1) reveal the link between Newton-discovered classical orbital velocity "V" along with orbital radius "R" of every captive of the same captor and their shared quantum-

originated orbital quantizer "Dc".

One can conclude that

> Nature has invented the concept of orbital quantizer "Dc", previously revealed via the equation (s1) in order to generate actual orbital velocities and orbital radii for all captives of the same captor. And this part of Nature's cosmic laws can be called here as the "orbital velocity-radius quantizer mechanism".

Equation Linking Orbital Radius Of A Captive To Its Orbital Quantizer:

Based on my previous discovery,

> The actual orbital radius of every captive must be quantumly created by its captor via the said orbital velocity-radius quantizer mechanism.

Orbital Radius Quantizer Equation:

The said orbital velocity-radius quantizer mechanism yields a specific equation to determine the actual orbital radius of a captive. This equation is deduced directly from the equation (x1) or its derived (x4) as:

$$R = R_c \times D_c^2 \tag{y1}$$

> where:
>
> R is the actual orbital radius of the captive with respect to its captor
>
> R_c is the quantum core radius, previously defined by (w2)
>
> D_c is the orbital quantizer generated by the captor for this captive

And in terms of concrete value:

$$R = [GM/c^2] \times D_c^2 \tag{y2}$$

The relationship between the calculated actual orbital radius of a captive, the quantum core radius (from its captor) and the orbital quantizer (for the captive) inside the equation (y2) can be visualized by the following diagram:

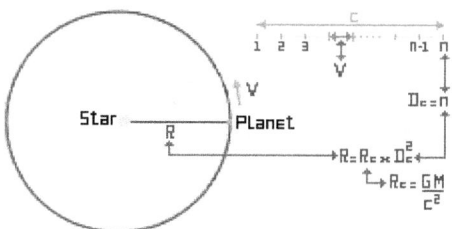

Connection Between Combination Of These Two Equations And Newton's Law Of Orbital Velocity Of Captives:

One can see that

The combination of the two previously postulated equations:

$(V \times Dc)^2 = GM / Rc$ (=x1)

$R = Rc \times Dc^2$ (=y1)

clearly leads to Newton's classical law of orbital velocity of captives, whose equation is:

$V^2 = GM/R$ (=u1)

as the three equations possess the same value of R and GM.

By the same token,

Newton's law of orbital motion of captives - presented here via its equation - appears to be an emergent law because it hides in plain sight a deeper quantum mechanism.

Born Rule Revealed By Orbital Quantizer Mechanism On Orbital Radius Of Captives

Born Rule In A Nutshell

The Born rule is known as

> A fundamental principle in quantum mechanics. The Born rule mathematically describes a quantum system that can yield the probabilities of various outcomes whenever "measurements" are performed on it.
>
> Historically, the Born rule was equationted by German physicist Max Born in 1926.

The Born rule states that,

> In a repeated system that comprises some state vector ψ and measures its overlap with another state vector ϕ, the fraction of measurements that find the system in state ϕ is equal to $|\langle\psi|\phi\rangle|^2$ (= has the probability proportional to the square of the amplitude).

Based on my assessment,

> The equation of the orbital velocity-radius-period-precession quantizer mechanism for captives dedicated to the orbital radius:
>
> $R = [GM/c^2] \times Dc^2$ (=y2)
>
> > where:
> > R is the orbital radius of a specific captive.
> > M is the mass of the captor.
> > Dc is the orbital quantizer created by the captor for a specific captive.
> > G is Newton's gravitational constant
> > c is the speed of light in vacuum
>
> appear to obey the Born rule.
>
> Should it be the case,
>
> > State vector ψ here is "GM/c^2" which is the gravitational energy

quanta.

State vector ɸ here is the orbital quantizer Dc.

The squared value of the state vector ɸ here – as "Dc^2" – determines physically the radial position of a captive with respect to its captor. And the said physical determination of the radial position of a captive happens when the wavefunction of the captor-captive pair's along its shared quantum axle – hence orbital radius of the captive – collapses.

Such collapse of the "radial" wavefunction is no other than the "measurement".

Captives' Motions Mirroring Wave-Particle Duality

Also based on my assessment,

Every captive appears to manifest in the Wave-particle duality with its perpetual orbital motion around its captor.

This is because:

Every captive is a physical object, therefore manifests itself as a "particle", albeit a giant one.

And at the same time operates a wave-like function that appear to follow the Born rule via the cube value of the assigned orbital quantizer. This wave-like function is described by the discovered orbital period quantizer equations:

$$T \times c / 2\pi = [GM/c^2] \times Dc^3 \qquad (=j1)$$

$$T = 2\pi \times ([GM/c^3] \times Dc^3) \qquad (=j2)$$

This wave-particle duality conforms with the Schrödinger equation paradigm via the orbital quantizer mechanism in the sense that

For each orbital quantizer sampling – which is same as an orbital frequency sampling – there are two possible quantum states for the physical captive:

Dead: In this quantum state, the physical captive's orbital motion is not activated

Alive: In this quantum state, the physical captive's orbital motion is activated.

The physical orbital velocity of the physical captive is the manifestation of the constant switching between the dead and alive quantum states via the captive's assigned orbital quantizer "Dc".

During one orbital quantizer/frequency sampling, the physical captive becomes alive only in one frequency beat then remains dead for all the remaining beats thereof.

The quantum system here is the determination of the radial and orbital location a captive with respect to its captor.

The various outcomes of this quantum system are the the radial location and the orbital location of a captive with respect to its captor.

The "measurements" here are whenever the orbital and radial wavefunctions of the captive collide with each other during the orbital motion of the said captive around its captor.

The phenomenon of perihelion precession anomaly of captives must be a major manifestation of Collapses of their Wavefunctions.

When the two said wave-fucnctions collide at one point, the clash must cause something. And that something may be no other than the perihelion precession anomaly (the most famous one is Mercury) Hence my discovered $\varepsilon = 6\pi \times (v/c)^2 / 1-e)^2$.

And 6π is same as $3 \times 2\pi$.

This quanta value of 3 inside my discovered perihelion precession anomaly equation also reveals the presence of the vibration wave-function of the gravitational field $2GM/c^2$.

Kepler's Second Law And Its Link With Discovered Orbital Period Quantizer Mechanism And Speed Of Gravity

Kepler's Second Law In A Nutshell

Kepler's second law states that

A planet moves in its orbit around the sun, it will sweep out equal areas in equal times.

The following diagram shows two orbital areas (Area 1 & Area 2) in wedge shape of different radial lengths but equal surfaces, and these areas are swept by the same planet during any two identical orbital periods (T1 & T2):

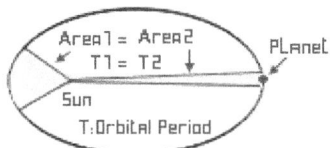

Gravitational Energy Spreading By Captor-Captive Quantum Axle As Quantum Mechanism Behind Kepler's Second Law

Term "[GM/c3] x Dc3" That Determines Orbital Period Of Captives Happens To Yield A Constant Value:

Based on my analysis,

My discovered equation of the orbital period of captives previously defined as:

$$T = 2\pi \times ([GM/c^3] \times Dc^3) \qquad (=j2)$$

shows that

The term "$[GM/c^3] \times Dc^3$" inside this equation remains constant during the whole orbital period of the captive, despite the fact that during this time frame the captive can find itself with ever-changing orbital velocity, in ever-changing location along the orbit and in any measured orbital period sampling.

This is because

> The generic gravitational mass value of the captor "GM" does not change over time.
>
> The gravitational energy quanta of the captor "GM/c^3" that is dedicated to the determination of the orbital period of any captive does not change over time either.
>
> The orbital quantizer "Dc" attributed to any captive by their captor does not change as it is calculated based on the mean value of all observed orbital velocities during the orbital period.

Evidence Of Quantum Origin Of Highest Orbital Velocity Of Captives At Perihelion And Lowest One At Aphelion:

There is an important fact that

> The orbital velocity of captives gradually changes between the aphelion (slowest) and the perihelion (fastest).

Then the question becomes:

> How does the captor and the captive quantumly manage in order in order to show up in our telescopic observation as Kepler's second law?

Based on my findings,

> The quantum nature of Kepler's second law could only work if there exists:
>
> > A quantum axle that connects at all times the captor with each specific captive.
> >
> > A quantum shuttle (structure of faster-than-speed-of-light particles?) that moves back and forth between the captor and each specific captive.
> >
> > This quantum shuttle receives a fixed gravitational energy quanta of the captor for each temporal sampling to spend (i.e. $1 GM/c^3$ for one second per SI standard).

The following diagram shows a quanta of gravitational energy that the said quantum shuttle receives and spends during an orbital period sampling of 4 seconds (n to n+3), and this quanta of gravitational energy is materialized by the total surface of a stack of rows (rectangles) filled each by a gravitational energy quanta portion (value of 1 x GM/c^3):

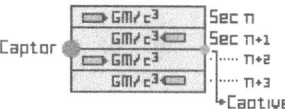

During this sampling of orbital period, the said quantum shuttle must travel back and forth between the captor and the captive, and must fully exhaust the received gravitational energy amount - represented by the said energy-filled total surface - regardless of the radial distances between the captor and the captive at any moment therein.

If one follows this quantum mechanism then,

When the captive is much closer to the captor, the said same gravitational energy quanta of the captor allows the said quantum shuttle to hit the captive many more times than when the captive is far away from the captor.

The following diagram shows how a sampling set of 4 average shuttle hits is equal to a set of 6 shuttle hits when the captor and captive are closer to each other and equal to a set of only 2 shuttle hits when the captor and captive are farther away from each other:

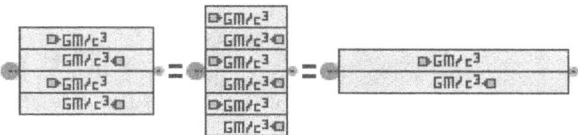

The subsidiary consequence of this quantum shuttle hit mechanism is that,

Every time the quantum shuttle hits the captive it forces the latter to advance along its orbit around the captor by an equal amount of distance.

And because of this quantum mechanism

With the same sampling of gravitational energy quanta (i.e. same "n x GM/c^3") produced during the same temporal sampling (i.e. same "n x seconds"),

> The bigger the number of times the quantum shuttle hits the captive, the longer the travel distance the captive will make along its orbit.
>
>> This quantum outcome gets translated into classical mechanics as a high velocity of the captive when the latter is close to the captor.
>
> On the contrary, the smaller the number of said captive-captor hits the quantum shuttle makes, the shorter the travel distance the captive makes along its orbit.
>
>> This quantum outcome gets translated into classical mechanics as a lower velocity of the captive when the latter is farther away from the captor.
>
> These two quantum scenarios can explain why a captive (planet or moon) reaches its highest orbital velocity at the perihelion and its slowest orbital velocity at the aphelion on its orbit around its captor (star or planet respectively).
>
>> These two quantum scenarios can also explain why the wedge-like orbital surfaces (as shown below) swept by a captive during the same temporal sampling is always the same regardless of where the captive is along its orbit and regardless of how fast or slow it moves.

● = Captor   ● = Captive   ←→ = GM/c^3

>> In the diagram above, the captive jumps from one position to the next along its orbit around the captor. The number of said jumps determines the actual orbital velocity value of the captive at that orbital segment.
>
>> The said wedge-like orbital surface generated by a moving captive is just the physical manifestation, hence the revelation, of the gravitational energy amount (equal to any number of quanta thereof) because the said wedge-like shape can be compressed or expanded along the quantum axle that connects the captor and the captive, but always its surface remains the same.
>
> These two quantum scenarios demonstrate that the equation of orbital

quantizer mechanism (j2) obeys the energy conservation law: any amount of gravitational energy that a captive receives from its captor must be fully spent regardless of period samplings.

The said quantum shuttle can be assimilated to a quantum twin that a physical captive is paired up with, and this quantum twin goes back and forth between the captor and the physical captive.

> The captive-owned quantum twin plays the role of spreading the gravitational energy that it receives from the captor.

In other words, the equation of orbital quantizer mechanism (j2) clearly demonstrates the deep quantum mechanism of Kepler's second law.

Quantum Origin Of Highest Orbital Velocity Of Captives At Perihelion And Lowest One At Aphelion Appears To Portray Heisenberg's Uncertainty Principle (Quantum Superposition And Wavefunction):

The described quantum mechanism behind Kepler's second law appears to demonstrate that there exists a dual mirroring between quantum mechanics (Heisenberg's uncertainty principle) and classical mechanics (Kepler's second law) and vice versa. This is how the mirroring gets materialized:

Heisenberg's uncertainty principle posits basically that:

> The position and the velocity/momentum of a particle cannot both be measured exactly, at the same time, as the particle itself is defined by it's wavefunction.

Werner Heisenberg – a German theoretical physicist – was one of the founders of quantum mechanics. His paper on the uncertainty principle, published in 1927, has become a foundational pillar of quantum mechanics.

Heisenberg's uncertainty principle is mathematically defined by the following equation:

$\Delta x \, \Delta p \geq \hbar/2$

> where
> Δx is the position of the particle
> Δp is the momentum of the particle
> $\hbar$ is the reduced Planck constant

In this particular quantum gravity mechanism, the captor and the captive behave like two celestial particles operating as two separate wavefunctions:

> The two celestial particles must be perpetually in a superposition of two eigenstates: immobile (dead) and mobile (alive). As a result, their position and momentum cannot be known until they are measured, aka observed.

The said quantum shuttle plays the role of a quantum measurement instrument. Each hit made by the quantum shuttle with the captor or the captive is no other than a measurement of the said celestial particles. At each hit moment, the wavefunction of the involved celestial particle collapses into a single eigenstate, and at this moment the quantum state thereof becomes "known".

When the celestial particle (captor or captive) is measured by the said quantum shuttle, its new quantum eigenstate (dead or alive) gets translated into the switch to a new classical state of orbital motion (immobile or mobile) of the said celestial particle, hence the change of the latter's momentum or position.

The described quantum mechanism behind Kepler's second law appears to also explain why both the captor and the captive rotate around their own axis as both of them are alternatively and repeatedly "measured" in their quantum world.

Speed Of Gravity Beyond Speed Of Light Revealed By Quantum Mechanism Behind Kepler's Second Law

Based on my analysis,

My discovered equation of the orbital period of captives previously defined as:

$$T = 2\pi \times ([GM/c^3] \times Dc^3) \qquad (=j2)$$

Reveals that gravity must have a speed based on the cube value of the speed of light (denoted as "c^3").

which would make the speed of gravity roughly 90 quadrillions (or 90,000 trillions) times faster than the speed of light.

This is because:

> The equation of the orbital period of captives (j2) can be rewritten as:

$$T / 2\pi = [GM/c^3] \times Dc^3 \qquad (j6)$$

$$T = 2\pi \times GM ([1/c^3] \times Dc^3) \qquad (j6.1)$$

$$T / 2\pi = GM ([1/c^3] \times Dc^3) \qquad (j6.2)$$

$$T / 2\pi = GM \times Dc^3 [1/c^3] \qquad (j6.3)$$

Let's examine a representative and simple sampling of orbital quantizer (Dc) with

$Dc = 299{,}792{,}458$ or orbital velocity of 1m/sec

In this sampling, (j6.3) becomes:

$$T/2\pi = GM \times 299{,}792{,}458^3 \times [1/299{,}792{,}458^3] = \quad \rightarrow$$

$$T/2\pi = GM \qquad (j6.4)$$

The resulting equation (j6.4) of the orbital period of captives demonstrates that

> The value ($T/2\pi$) of the radial period of back-and-forth communication between the captor and the captive when the latter moves with an orbital velocity of 1m/second must be equal to the gravitational mass of the captor (1GM).

And

> The said back-and-forth communication between the captor and the captive, which is no other than the speed of gravity, must operate with the speed equal to the cube value (c^3) of the speed of light.

This special case confirms that the speed of gravity can reach the cube value of the speed of light.

Let's examine a representative and simple sampling of orbital quantizer (Dc) with

$Dc = 149{,}896{,}229$ (or $299{,}792{,}458/2$ or $c/2$) or orbital velocity of 2m/sec

In this sampling, (j6.3) becomes:

$$T/2\pi = GM \times Dc^3 [1/c^3] \qquad (=j6.3)$$

$$T/2\pi = GM \times (c/2)^3 \times [1/c^3] \qquad \rightarrow$$

$$T/2\pi = GM [(c^3/2^3)/c^3] \qquad \rightarrow$$

$$T/2\pi = GM [c^3/c^3] \times 1/2^3 \qquad (j6.5)$$

The resulting equation (j6.5) of the orbital period of captives demonstrates that

> The value ($T/2\pi$) of the radial period of back-and-forth communication between the captor and the captive when the latter moves with an orbital velocity of 2m/second must be equal to the gravitational mass of the captor ($1GM/2^3$).

And

> The said back-and-forth communication between the captor and the captive, which is no other than the speed of gravity, must also operate with the speed equal to the cube value (c^3) of the speed of light, despite a shorter orbital period by a factor of 8 ($=2^3$).

These special cases reveal that,

The orbital period is reduced by a factor of cube value of the reduction ratio of the orbital quantizer of a given captive whenever the latter changes.

How Is This Discovered Speed Of Gravity "c^3" Possible When LIGO-Detected Gravitational Waves Do Not Surpass The Speed Of Light?

Based on my analysis,

> The speed of gravity's actions can be very well different from the speed of gravity's artifacts.

By the same token,

> Gravity's actions are not matter like gravitational waves which are artifacts thereof, therefore are not constrained by the limit of motion of matter, discovered to be equal to the speed of light.

And there is one more reason:

> There has already been one major discovery in physics that portrays a similar pattern: the quantum entanglement.
>
>> The actions of quantum entanglement appear to be non-local, or "instantaneous", and yet the artifacts thereof still obey the physical constraint of the speed of light.

Kepler's Third Law And Its Link With Discovered Orbital Period-Radius-Velocity Quantizer Mechanism

Kepler's Third Law In A Nutshell

Kepler's third law states that

> The square value of the orbital period of any planet is proportional to the cube value of the semi-major axis of its orbit.

Newton's Version Of Kepler's Third Law:

Newton had created his own version of Kepler's third law via the following equation (t0):

$$T^2 = \frac{4\pi^2}{G(M_1+M_2)} a^3$$

which is same as:

$\qquad T^2 = [4\pi^2 / G(M_1+M_2)] \times (R^3)$ \hfill (t0)

where:
T is the orbital period of the captive.

a is the semi-major axis of the elliptical orbit of the captive.
G is Newton's gravitational constant
M_1 is the mass of the captor
M_2 is the mass of the captive

In the case where the captor's mass is highly dominant (like the Sun in the Solar system) then M_2 can be ignored, hence a shorter Newton version of Kepler's third law as follows:

$\qquad T^2 = [4\pi^2 / GM] \times (R^3)$ \hfill (j2.7)

Gravitational Energy Spreading By Captor-Captive Quantum Axle As Quantum Mechanism Behind Kepler's Third Law

Based on my findings,

The equation of orbital quantizer mechanism (j2), as shown hereafter, reveals the deep quantum mechanism of Kepler's third law.

$$T = 2\pi \times ([GM/c^3] \times Dc^3) \quad (=j2)$$

Here is how the deep quantum mechanism of Kepler's third law works:

First, by equally raising the left-hand and right-hand sides of the equation of orbital quantizer mechanism (j2) to the power of 2, we get:

$$T^2 = [2\pi \times ([GM/c^3] \times Dc^3)]^2 \quad (j2.1)$$

Then by integrating the exponent 2 with each term in the new transitional equation, we get:

$$T^2 = 4\pi^2 \times ([(GM)^2/c^6] \times Dc^6) \quad (j2.2)$$

Because the exponent of 6 is same as the exponent of 2x3, then we get:

$$T^2 = 4\pi^2 \times ([(GM)^2/c^{2\times 3}] \times Dc^{2\times 3}) \quad (j2.3)$$

As GM/GM = 1, we can multiply the right-hand side of the transitional equation above by "GM/GM" without changing its value in order to obtain a new transitional equation as:

$$T^2 = [4\pi^2 / GM] \times ([GM \times (GM)^2/c^{2\times 3}] \times Dc^{2\times 3}) \quad (j2.4)$$

which turns out to be identical to:

$$T^2 = [4\pi^2 / GM] \times ([(GM)^3/c^{2\times 3}] \times Dc^{2\times 3}) \quad (j2.5)$$

As the terms GM, c^2 and Dc^2 share the same exponent of 3 then we can set the latter as their common exponent, hence:

$$T^2 = [4\pi^2 / GM] \times ([GM/c^2] \times Dc^2)^3 \quad (j2.6)$$

It turns out that the element $([GM/c^2] \times Dc^2)$ has a quantum function of its own. This quantum function was revealed by my previous discovered equation:

$$R = [GM/c^2] \times Dc^2 \quad (=y2)$$

where

R is the orbital radius (aka semi-major axis) of the captive

This quantum function along with its equation defines how an orbital radius and orbital velocity is generated by a captor for each of its captives. Their full explanation is presented in my book:

> Orbital Velocity-Radius Quantizer Mechanism Hidden Behind Newton-Einstein Gravity

Therefore,

$$T^2 = [4\pi^2 / GM] \times (R^3) \qquad (=j2.7)$$

This final transitional equation (j2.7) shows that

> The square value of the orbital period of the captive is proportional to the cubic value of its orbital radius (aka semi-major axis), which is exactly the postulate of Kepler's third law.

Furthermore, this final transitional equation (j2.7) shows that it also matches Newton's version of Kepler's third law.

Orbital Velocity-Radius Quantizer Mechanism Behind Newton's Law Of Orbital Velocity Of Celestial Objects

Based on my previous findings,

> The actual orbital radius and the actual orbital velocity of each captive with respect to its captor are not random values, but are the outcome values of a cosmic mechanism with quantum characteristics.

And I have written a book to describe it, and

> The related book's name is:
>
> > Orbital Velocity-Radius Quantizer Mechanism Hidden Behind Newton-Einstein Gravity

Reason-For-Being Of A Hidden Orbital Velocity-Radius Quantizer Mechanism

Based on my findings, there appears to exist

> A quantum cosmic mechanism that every captor uses to specify the orbital radius at which each of its captives can stay in coordination with a specific orbital velocity.

Such a quantum cosmic mechanism appears to

> Allow a captor to harmonize the orbital motions of its captives in an orderly fashion at all cosmic levels. Without it, there would be a set of chaotic orbits and subsequent risks of random collisions between captives at multiple levels: a cosmic scenario that has not been observed so far.

It is possible that nature has the possibility to create different cosmic mechanisms with the same purpose, but my findings show that there is one that appears to be quantum-friendly: simple mechanism via increment of discrete values.

Internal Core Functions Of Orbital Velocity-Radius Quantizer Mechanism

There are two core functions that constitute the orbital velocity-radius quantizer mechanism.

These two core functions are to define:

The suitable orbital quantizer Dc via the equation:

$$Dc = c / V \qquad (=s1)$$

The suitable orbital radius R via the equation:

$$R = Rc \times Dc^2 \qquad (=y1)$$

where

$$Rc = GM/c^2 \qquad (=w2)$$

These two above functions go hand in hand because if there is any change of value in one function equation, automatically it will trigger the change of value in the other.

Furthermore, these two above equations are mathematically reciprocal. Therefore the two functions are interchangeable. Concretely, it means that:

The captor can trigger the function to determine the suitable quantizer Dc via the speed of light first, then automatically deduce the suitable orbital velocity and finally the suitable orbital radius.

or

The captor can trigger the function to determine the suitable orbital quantizer Dc via the gravitational energy potential quanta of the captor first, then automatically deduce the suitable orbital radius and finally the suitable orbital velocity.

In other words, no matter which said function is used by the captor to determine the suitable orbital quantizer, the found value thereof must be the same for both the said functions when it comes to the application of the orbital velocity-radius quantizer mechanism.

Function Of Determination Of Orbital Quantizer Via Speed Of Light To Obtain Gravitationally Suitable Radius Of A Captive

Based on my previous findings,

The discovered equations (s1), (w2), (y1) and (v1), as shown below, lead to the revelation of a quantum mechanism behind Newton's classical equation of orbital velocity of captives.

$$Dc = c / V \qquad (=s1)$$

$$Rc = GM/c^2 \qquad (=w2)$$

$$R = Rc \times Dc^2 \qquad (=y1)$$

$$\rightarrow$$

$$V^2 = GM / [Rc \times Dc^2] \qquad (=v1)$$

$$\rightarrow$$

$$V^2 = GM/R \qquad (=u1)$$

The ultimate role of this orbital velocity-radius quantizer mechanism must be to:

Help the captor to Determine the gravitationally safe orbital radius of a captive.

The notion of gravitational safety of a captive's orbital radius here means that,

This gravitationally safe orbital radius must provide an orbital setting that allows the involved captive to revolve around the captor without risks of colliding with other captives.

This gravitational safety is possible because each captive has a unique orbital quantizer, unless it is part of a chain of co-orbital captives that share the same orbit.

This gravitational safety is one of the suitability criteria that determine in the end the exact radial location of the involved captive.

Function Of Determination Of Orbital Quantizer Via Captor's Gravitational Energy Potential To Obtain Gravitationally Suitable Radius Of A Captive

Based on my previous findings,

Each captor must possess a quantum function to determine the value of the orbital quantizer for each of its captives by means of its gravitational energy potential quanta in order to obtain a gravitationally suitable orbital radius of a captive and also the latter's orbital velocity as a consequence.

The classical orbital radius outcome (R) of this quantum function is defined via the discovered equation:

$$R = R_c \times D_c^2 \qquad (=y1)$$

or in its full expanded equation:

$$R = [GM/c^2] \times D_c^2 \qquad (=y2)$$

Quantum Function's Components:

By the same token, this quantum function has two components:

The quantum core radius for all captives (R_c).

The orbital quantizer for a specific captive (D_c).

Quantum Function's Simple Mathematical Operation:

This quantum function operates a:

Simple multiplication operation with two components:

The quantum core radius (R_c) is the multiplicand.

The square value of the orbital quantizer (D_c) is the multiplier.

Evidence Of Orbital Velocity-Radius Quantizer Mechanism Through Planets And Asteroids Of The Sun

Evidence Of Orbital Velocity-Radius Quantizer Mechanism Through Planets of The Solar System

Targets Of This Orbital Velocity-Radius Quantizer Mechanism Verification:

Based on my calculations, there appears to exist a host of evidence of this orbital velocity-radius quantizer mechanism for all the known planets inside the Solar system. The planets subjected to this verification are:

Mercury, Venus, Earth. Mars, Jupiter, Saturn, Uranus, Neptune

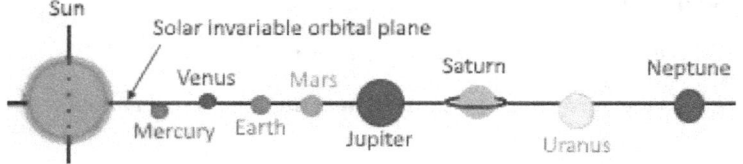

This basis of this segment has been fully described in my book:

Orbital Velocity-Radius Quantizer Mechanism Hidden Behind Newton-Einstein Gravity

As a result, only a summary thereof will be presented hereafter in order to show

The fundamental link between the three types of orbital quantizers that appear to come from a potential discovered quantum gravity law for every captive with respect to its captor: the orbital velocity, orbital radius and orbital period.

Purposes Of This Orbital Velocity-Radius Verification:

The first purpose of this verification is to prove

The existence of the orbital quantizer (denoted as Dc) for orbital velocities of captives via the equation:

$$Dc = c / V \qquad (=s1)$$

where
c is the speed of light
V is the actual orbital velocity of a given captive (like planet with respect to the Sun)

The second purpose of this verification is to prove

The existence of the orbital quantizer (denoted as Dc) for orbital radii of captives via the equation:

$$R = [GM/c^2] \times Dc^2 \qquad (=y2)$$

Based on my findings,

The (y2) equation can be deduced from Newton's law equation of orbital velocity of a captive:

$$V^2 = GM / R$$

Which leads to:

$$R = GM / V^2$$

And because from (s1):

$$V = c / Dc$$

Then:

$$R = GM / V^2 = GM / (c / Dc)^2$$

which is same as:

$$R = [GM/c^2] \times Dc^2 \qquad (=y2)$$
$$R = Rc \times Dc^2 \qquad (=y1)$$

The equation (y2) of the orbital radius of a captive shows that the latter is solely based on the value of the intrinsic gravitational energy potential quanta Eg_1 (GM/c^2) of the Sun and the square value of the orbital quantizer (Dc) of the said captive.

Reference Data Of This Orbital Velocity-Radius Verification: Of Sun's Captives (Planets, Planetoids And Asteroids):

The reference data here is:

Sun's Gravitational Energy Value 1GM/c²

The Sun's effective gravitational energy potential quanta dedicated exclusively to orbital motion of each captive (E_{g1}) as:

GM_{sun}/c^2 = 1,476.6919 m or 1.476619 km

based on the values of:

M_{sun} = 1.9885e+30 kg (as mass of the Sun)
G = 6.6743e-11 (N kg^{-2} m²)
c = 299,792,458 m

Let's go through each planet subjected to this verification.

Orbital Velocity-Radius Verification Case of Mercury:

Observed astronomical data of orbital period and orbital circumference of Mercury yields:

Mercury's observed orbital velocity value as:

$V_{mercury}$ = 47.87 km/s or 47,870 m/s

Then the equation (s1) yields Mercury's orbital quantizer D_c value as:

$D_{c\_mercury}$	= c / $V_{mercury}$
	= 299,792,458 m / 47,870 m →
$D_{c\_mercury}$	= 6,262.62

Therefore Mercury's deduced radius $R_{mercury}$ becomes:

$R_{mercury}$ =	= [GM_{sun}/c^2] x $D_{c\_mercury}^2$
	= 1,476.6919 m x 6,262.62² →
$R_{mercury}$	= 57,916,386,156 m or 57,916,386 km

which tightly matches Mercury's observed orbital radius value as: 57,910,000 km (Δ=0.011%)

Orbital Velocity-Radius Verification Case of Venus:

And with Venus' observed orbital velocity value as:

V_{venus} = 35.02 km/s or 35,020 m/s

Venus' orbital quantizer D_{c_venus} value becomes:

D_{c_venus} = c / V_{venus}

	= 299,792,458 m / 35,020 m	→
Dc_venus	= 8,560.60	

Therefore Venus' variable radius Rvenus becomes:

Rvenus	= [GMsun/c²] x Dc_venus²	
	= 1,476.6919 m x 8,560.60²	→
Rvenus	= 108,217,700,714 m or 108,217,700 km	

which tightly matches Venus' observed orbital radius value as: 108,200,000 km (Δ=0.016%) or 108,209,475 according to NASA (Δ=0.0077%)

Orbital Velocity-Radius Verification Case of Earth:

With Earth's observed orbital velocity value as:

Vearth = 29.78 km/s or 29,780 m/s

Earth's orbital quantizer Dcearth value becomes:

Dcearth	= c / Vearth	
	= 299,792,458 m / 29,780 m	→
Dcearth	= 10,066.90	

Therefore Earth's variable radius Rearth becomes:

Rearth	= [GMsun/c²] x Dc_earth²	
	= 1,476.6919 m x 10,066.90²	→
Rearth	= 149,651,612,859 m or 149,651,612 km	

which tightly matches Earth's observed orbital radius value as: 149,598,000 km (Δ=0.035%)

Orbital Velocity-Radius Verification Case of Mars:

With Mars' observed orbital velocity value as:

Vmars = 24.077 km/s or 24,077 m/s

Mars' orbital quantizer Dc_mars value becomes:

Dc_mars	= c / Vmars	
	= 299,792,458 m / 24,077 m	→
Dc_mars	= 12,451.40	

Therefore Mars' variable radius R$_{mars}$ becomes:

R$_{mars}$ = [GMsun/c^2] x Dc_mars^2
= 1,476.6919 m x 12,451.40^2 →

R$_{mars}$ = 228,942,416,603 m or 228,942,416 km

which tightly matches Mars' observed orbital radius value as: 227,936,637 km (Δ=0.441%)

Orbital Velocity-Radius Verification Case of Mathilde:

With Mathilde's observed orbital velocity value as:

V$_{mathilde}$ = 17.98 km/s or 17,980 m/s

Mathilde's orbital quantizer Dc_mathilde value becomes:

Dc_mathilde = c / V$_{mathilde}$
= 299,792,458 m / 17,980 m   →

Dc_mathilde = 16,673.66

Therefore Mathilde's variable radius R$_{mathilde}$ becomes:

R$_{mathilde}$ = [GMsun/c^2] x Dc_mathilde^2
= 1,476.6919 m x 16,673.66^2 →

R$_{mathilde}$ = 410,536,499,954 m or 410,536,499 km

which tightly matches Mathilde's observed orbital radius value as: 396,200,961 km (Δ=03.6182%)

Orbital Velocity-Radius Verification Case of Juno:

With Juno's observed orbital velocity value as:

V$_{juno}$ = 17.93 km/s or 17,930 m/s

Juno's orbital quantizer Dc_juno value becomes:

Dc_juno = c / V$_{juno}$
= 299,792,458 m / 17,930 m   →

Dc_juno = 16,720.16

Therefore Juno's variable radius R$_{juno}$ becomes:

R$_{juno}$ = [GMsun/c^2] x Dc_juno^2

R_{juno} = 1,476.6919 m x 16,720.16² →

R_{juno} = 412,829,525,787 m or 412,829,525 km

which tightly matches Juno's observed orbital radius value as: 399,536,720 km (Δ=03.327%)

Orbital Velocity-Radius Verification Case of Eugenia:

With Eugenia's observed orbital velocity value as:

$V_{eugenia}$ = 18.06 km/s or 18,060 m/s

Eugenia's orbital quantizer $Dc\_eugenia$ value becomes:

$Dc\_eugenia$ = c / $V_{eugenia}$
= 299,792,458 m / 18,060 m →

$Dc\_eugenia$ = 16,599.80

Therefore Eugenia's variable radius $R_{eugenia}$ becomes:

$R_{eugenia}$ = [GM_{sun}/c^2] x $Dc\_eugenia^2$
= 1,476.6919 m x 16,599.80² →

$R_{eugenia}$ = 406,907,414,788 m or 406,907,414 km

which tightly matches Eugenia's observed orbital radius value as: 406,897,040 km (Δ=0.0025%)

Orbital Velocity-Radius Verification Case of Ceres:

With Ceres' observed orbital velocity value as:

V_{ceres} = 17.87 km/s or 17,870 m/s

Ceres' orbital quantizer Dc_ceres value becomes:

Dc_ceres = c / V_{ceres}
= 299,792,458 m / 17,870 m →

Dc_ceres = 16,776.29

Therefore Ceres' variable radius R_{ceres} becomes:

R_{ceres} = [GM_{sun}/c^2] x Dc_ceres^2
= 1,476.6919 m x 16,776.29² →

R_{ceres} = 415,605,936,536 m or 415,605,936 km

which tightly matches Ceres' observed orbital radius value as: 414,092,800 km ($\Delta=0.0293\%$)

Orbital Velocity-Radius Verification Case of Pallas:

Pallas is the second largest of the four largest planetoids (Ceres, Pallas, Juno, Vesta) inside the Solar system's Main Asteroid Belt.

With Pallas' observed orbital velocity value as:

V_{pallas} = 17.88 km/s or 17,880 m/s
(value calculated from orbital radius ans period)

It is important to notice that the public data reported value of V_{pallas} is 17.94 km/s, which is faster than that of Ceres, and yet Pallas is further away from the Sun, therefore inconsistent and cannot be seen as reliable

Pallas' orbital quantizer Dc_{pallas} value becomes:

Dc_{pallas} = c / V_{pallas}
= 299,792,458 m / 17,880 m →

Dc_{pallas} = 16,766.91

Therefore Pallas' variable radius R_{pallas} becomes:

R_{pallas} = $[GM_{sun}/c^2] \times Dc\_{pallas}^2$
= 1,476.6919 m × 16,766.91² →

R_{pallas} = 415,141,317,261 m or 415,141,317 km

which tightly matches Pallas' observed orbital radius value as: 414,691,200 km ($\Delta=0.1085\%$)

Orbital Velocity-Radius Verification Case of Jupiter:

With Jupiter's observed orbital velocity value as:

$V_{jupiter}$ = 13.07 km/s or 13,070 m/s

Jupiter's orbital quantizer $Dc\_{jupiter}$ value becomes:

$Dc\_{jupiter}$ = $c / V_{jupiter}$
= 299,792,458 m / 13,070 m →

$Dc\_{jupiter}$ = 22,937.44

Therefore Jupiter's variable radius $R_{jupiter}$ becomes:

$R_{jupiter}$ = $[GM_{sun}/c^2]$ x $Dc\_jupiter^2$
= 1,476.6919 m x 22,937.44² →

$R_{jupiter}$ = 776,926,229,626 m or 776,926,229 km

which tightly matches Jupiter's observed orbital radius value as: 778,412,027 km (Δ=0.19%)

Orbital Velocity-Radius Verification Case of Saturn:

With Saturn's observed orbital velocity value as:

V_{saturn} = 9.69 km/s or 9,690 m/s

Saturn's orbital quantizer Dc_saturn value becomes:

Dc_saturn = c / V_{saturn}
= 299,792,458 m / 9,690 m →

Dc_saturn = 30,938.33

Therefore Saturn's variable radius R_{saturn} becomes:

R_{saturn} = $[GM_{sun}/c^2]$ x Dc_saturn^2
= 1,476.6919 m x 30,938.33² →

R_{saturn} = 1,413,460,341,490 m or 1,413,460,341 km

which tightly matches Saturn's observed orbital radius value as: 1,426,725,400 km (Δ=0.93%)

Orbital Velocity-Radius Verification Case of Uranus:

With Uranus' observed orbital velocity value as:

V_{uranus} = 6.81 km/s or 6,810 m/s

Uranus' orbital quantizer Dc_uranus value becomes:

Dc_uranus = c / V_{uranus}
= 299,792,458 m / 6,810 m →

Dc_uranus = 44,022.38

Therefore Uranus' variable radius R_{uranus} becomes:

R_{uranus} = [GMsun/c²] x Dc_uranus²
= 1,476.6919 m x 44,022.38² →

R_{uranus} = 2,861,784,514,117 m or
2,861,784,514 km

which tightly matches Uranus' observed orbital radius value as: 2,870,972,219 km (Δ=0.32%)

Orbital Velocity-Radius Verification Case of Neptune:

With Neptune's observed orbital velocity value as:

$V_{neptune}$ = 5.43 km/s or 5,430 m/s

Neptune's orbital quantizer Dc_neptune value becomes:

Dc_neptune = c / Vneptune
= 299,792,458 m / 5,430 m →

Dc_neptune = 55,210.39

Therefore Neptune's variable radius Rneptune becomes:

$R_{neptune}$ = [GMsun/c²] x Dc_neptune²
= 1,476.6919 m x 55,210.39² →

$R_{neptune}$ = 4,501,233,294,692 m or
4,501,233,294 km

which tightly matches Neptune's observed orbital radius value as: 4,498,252,910 km (Δ=0.066%)

Orbital Velocity-Radius Verification Case of Pluto:

With Pluto's observed orbital velocity value as:

V_{pluto} = 4.74 km/s or 4,740 m/s

Pluto's orbital quantizer Dc_pluto value becomes:

Dc_pluto = c / Vpluto
= 299,792,458 m / 4,740 m →

Dc_pluto = 63,247.35

Therefore Pluto's variable radius Rpluto becomes:

R_{pluto} = [GMsun/c²] x Dc_pluto²
= 1,476.6919 m x 63,247.35² →

R_{pluto} = 5,907,103,225,521 m or
5,907,103,225 km

which tightly matches Pluto's observed orbital radius value as: 5,906,376,272 km (Δ=0.066%)

Orbital Velocity-Radius Verification Case of Eris:

With Eris's observed orbital velocity value as:

V_{eris} = 3.434 km/s or 3,434 m/s

Then Eris's orbital quantizer Dc_eris value becomes:

Dc_eris = c / Veris
= 299,792,458 m / 3,434 m →

Dc_eris = 87,301.24

Therefore Eris's variable radius Reris becomes:

R_{eris} = [GMsun/c²] x Dc_eris²
= 1,476.6919 m x 87,301.24² →

R_{eris} = 11,254,616,922,524 m or
11,254,616,922 km

which does not match Eris's observed orbital radius value as: 10,152,454,400 km (Δ=10.85%) at first hand

However, as Eris's orbital eccentricity (e) of 0.44 is very high, therefore we need to apply ax√(1-e²) as follows:

R_{eris} x √(1-e²) = 11,254,616,922 m x √(1-0.44²)
= 11,254,616,922 m x 0.89799 →

R_{eris} x √(1-e²) = 10,106,533,449 m

which matches Eris's observed orbital radius value as: 10,152,454,400 km (Δ=0.45%)

Orbital Velocity-Radius Verification Case of Sedna:

With Sedna's observed orbital velocity value as:

V_{sedna} = 1.04 km/s or 1,040 m/s

Sedna's orbital quantizer D_{c_sedna} value becomes:

D_{c_sedna} = c / V_{sedna}
= 299,792,458 m / 1,040 m →

D_{c_sedna} = 288,262

Therefore Sedna's variable radius R_{sedna} becomes:

R_{sedna} = [GM_{sun}/c^2] x $D_{c\_sedna}^2$
= 1,476.6919 m x $288,262^2$ →

R_{sedna} = 122,705,684,847,651 m or
122,705,684,847 km

which does not match Sedna's observed orbital radius value as: 75,697,600,000 km (Δ=62.09%) at first hand

However, as Sedna's orbital eccentricity (e) of 0.8496 is very high, therefore we need to apply $a \times \sqrt{1-e^2}$ as follows:

R_{sedna} x $\sqrt{1-e^2}$ = 122,705,684,847,651 m x $\sqrt{1-0.8496^2}$
= 122,705,684,847,651 m x 0.527427 →

R_{sedna} x $\sqrt{1-e^2}$ = 64,718,291,242,142 m

which matches Sedna's observed orbital radius value as: 75,697,600,000 km (Δ=16.96%)

This calculation appears to reveal that

Either Sedna's Orbital Velocity of 1.04 km/second or Sedna's Orbital radius of 75,697,600,000 km is not quite reliable.

Orbital Velocity-Radius Verification Case of The Goblin:

As of 2023, there is no available data about the observed orbital velocity of the Goblin. The only available data (from Theskylive.com) are:

The Goblin's Orbital Radius (From ~ 1,369.79770882 AU)	204,918,820,941 km
The Goblin's Perihelion Radius (From ~ 65.11260610 AU)	9,740,707,247 km
The Goblin's Aphelion Radius (From ~ 2,674.48281155 AU)	400,096,934,634 km
The Goblin's Orbital Period	40,000 Earth-years or

1.26144e+12 seconds

From these data, we can deduce

The Goblin's Orbital Velocity   1.020 km/s
(From 204,918,820,941 km x 2π /
1.26144e+12 seconds)

With The Goblin's observed orbital velocity value as:

V_{goblin} = 1.02 km/s or 1,020 m/s

The Goblin's orbital quantizer D_{c_goblin} value becomes:

D_{c_goblin} = c / V_{goblin}
                = 299,792,458 m / 1,020 m   →

D_{c_goblin} = 293,914

Therefore The Goblin's variable radius R_{goblin} becomes:

R_{goblin} = [GM_{sun}/c^2] x $D_{c\_goblin}^2$
           = 1,476.6919 m x 293,914^2 →

R_{goblin} = 127,564,678,634,014 m or
           127,564,678,634 km

which does not match the Goblin's observed orbital radius value as: 204,918,820,941 km (Δ=60.63%) at first hand

However, as the Goblin's orbital eccentricity (e) of 0.0 is very high, therefore we need to apply $a_x\sqrt{1-e^2}$ as follows:

R_{goblin} x $\sqrt{(1-e^2)}$ = 122,705,684,847,651 m x $\sqrt{(1-0.0^2)}$
                     = 122,705,684,847,651 m x 0.0   →

R_{goblin} x $\sqrt{(1-e^2)}$ = 0 m

which matches the Goblin's observed orbital radius value as: 204,918,820,941 km (Δ=0.0%)

This calculation appears to reveal that

The Goblin's Orbital Period of 40,000 Earth-years is not a reliable data. The rounded number appears already to be an approximated value.

Evidence Of Orbital Velocity-Radius Quantizer Mechanism Through Asteroids of The Sun

Based on my calculations, there also appears to exist a host of evidence of this orbital velocity-radius quantizer mechanism for a host of known asteroids that revolve around the Sun, just like the one for the latter's planetary counterparts. The asteroids subjected to this verification are:

Ra-shalom, Aten, Icarus, Phaethon, Toro, Apollo, Adonis

Orbital Velocity-Radius Verification Case of Ra-shalom:

From the astronomically

Observed mean orbital velocity of Ra-shalom ($V_{rashalom}$) as:

$V_{rashalom}$ = 32.652 km/s or 32,652 m/s
Source:

Ra-shalom's orbital quantizer $Dc\_rashalom$ value becomes:

$Dc\_rashalom$ = c / $V_{rashalom}$
= 299,792,458 m / 32,652 m   →

$Dc\_rashalom$ = 9,181.44

Therefore Ra-shalom's variable radius $R_{rashalom}$ becomes:

$R_{rashalom}$ = [GM_{sun}/c^2] x $Dc\_rashalom^2$
= 1,476.6919 m x 9,181.44²   →

$R_{rashalom}$ = 124,483,414,906 m or 124,483,414 km

which tightly matches Ra-shalom's observed orbital radius value as: 124,467,200 km (Δ=0.013%)

Orbital Velocity-Radius Verification Case of Aten:

From the astronomically

Observed mean orbital velocity of Aten (V_{aten}) as:

V_{aten} = 30.336 km/s or 30,336 m/s
Source:

Aten's orbital quantizer Dc_aten value becomes:

Dc_aten = c / V_{aten}
= 299,792,458 m / 30,336 m   →

Dc_aten = 9,882.39

Therefore Aten's variable radius R_{aten} becomes:

R_{aten} = $[GM_{sun}/c^2]$ x $D_{c\_aten}^2$
= 1,476.6919 m x 9,882.39² →

R_{aten} = 144,216,141,080 m or 144,216,141 km

which tightly matches Aten's observed orbital radius value as: 144,648,240 km (Δ=0.3%)

Orbital Velocity-Radius Verification Case of Icarus:

From the astronomically

Observed mean orbital velocity of Icarus (V_{icarus}) as:

V_{icarus} = 28.69 km/s or 28,690 m/s
Source:
https://www.spacereference.org/asteroid/1566-icarus-1949-ma
https://academickids.com/encyclopedia/index.php/1566_Icarus#google_vignette

Icarus' orbital quantizer D_c_{icarus} value becomes:

D_{c_icarus} = c / V_{icarus}
= 299,792,458 m / 28,690 m →

D_{c_icarus} = 10,449.37

Therefore Icarus' variable radius R_{icarus} becomes:

R_{icarus} = $[GM_{sun}/c^2]$ x $D_{c\_icarus}^2$
= 1,476.6919 m x 10,449.37² →

R_{icarus} = 161,239,004,193 m or 161,239,004 km

which tightly matches Icarus' observed orbital radius value as: 161,283,760 km (Δ=0.03%)

Orbital Velocity-Radius Verification Case of Phaethon:

From the astronomically

Observed mean orbital velocity of Phaethon ($V_{phaethon}$) as:

$V_{phaethon}$ = 26.415 km/s or 26,415 m/s
Source:

Phaethon's orbital quantizer $D_{c\_phaethon}$ value becomes:

$Dc\_phaethon$ = c / Vphaethon
= 299,792,458 m / 26,415 m →

$Dc\_phaethon$ = 11,349.32

Therefore Phaethon's variable radius Rphaethon becomes:

Rphaethon = $[GMsun/c^2]$ x $Dc\_phaethon^2$
= 1,476.6919 m x $11,349.32^2$ →

Rphaethon = 190,208,348,754 m or 190,208,348 km

which tightly matches Phaethon's observed orbital radius value as: 190,201,440 km (Δ=0.0036%)

Orbital Velocity-Radius Verification Case of Toro:

From the astronomically

Observed mean orbital velocity of Toro (Vtoro) as:

Vtoro = 25.471 km/s or 25,471 m/s
Source:

Toro's orbital quantizer Dc_toro value becomes:

Dc_toro = c / Vtoro
= 299,792,458 m / 25,471 m →

Dc_toro = 11,769.95

Therefore Toro's variable radius Rtoro becomes:

Rtoro = $[GMsun/c^2]$ x Dc_toro^2
= 1,476.6919 m x $11,769.95^2$ →

Rtoro = 204,568,673,250 m or 204,568,673 km

which tightly matches Toro's observed orbital radius value as: 204,578,000 km (Δ=0.005%)

Orbital Velocity-Radius Verification Case of Apollo:

From the astronomically

Observed mean orbital velocity of Apollo (Vapollo) as:

Vapollo = 24.552 km/s or 24,552 m/s
Source:

Apollo's orbital quantizer D_{c_apollo} value becomes:

D_{c_apollo}	= c / V_{apollo}
	= 299,792,458 m / 24,552 m →
D_{c_apollo}	= 12,210.51

Therefore Apollo's variable radius R_{apollo} becomes:

R_{apollo}	= [GM_{sun}/c^2] x $D_{c\_apollo}^2$
	= 1,476.6919 m x 12,210.51^2 →
R_{apollo}	= 220,169,674,289 m or 220,169,674 km

> which tightly matches Apollo's observed orbital radius value as: 219,963,671 km (Δ=0.093%)

Orbital Velocity-Radius Verification Case of Adonis:

From the astronomically

> Observed mean orbital velocity of Adonis (V_{adonis}) as:

V_{adonis} = 21.769 km/s or 21,769 m/s
Source:

Adonis' orbital quantizer D_{c_adonis} value becomes:

D_{c_adonis}	= c / V_{adonis}
	= 299,792,458 m / 21,769 m →
D_{c_adonis}	= 13,771.53

Therefore Adonis' variable radius R_{adonis} becomes:

R_{adonis}	= [GM_{sun}/c^2] x $D_{c\_adonis}^2$
	= 1,476.6919 m x 13,771.53^2 →
R_{adonis}	= 280,062,059,207 m or 280,062,059 km

> which tightly matches Adonis' observed orbital radius value as: 280,476,258 km (Δ=0.148%)

Evidence Of Orbital Velocity-Radius Quantizer Mechanism Through Earth's Moon And Satellite

Targets Of This Orbital Velocity-Radius Quantizer Mechanism Verification Of Earth's Moon And Satellites:

Based on my calculations, there appears to exist a host of evidence of this orbital velocity-radius quantizer mechanism for all the known natural and artificial moons and satellites inside the Earth's planetary system. The moons subjected to this verification are:

The Moon, artificial satellite LAGEOS-II

Orbital Velocity-Radius Verification Case of Earth's Moon:

From the equation (z1) we obtain:

The Earth's effective gravitational energy potential quanta dedicated exclusively to orbital motion of each captive (E_{g1}) as:

$GM_{earth}/c^2 = 0.0044351999$ m

based on values of:

$M_{earth} = 5.9724e+24$ kg (as mass of the Earth)
$G = 6.6743e-11$ (N kg^{-2} m^2)
$c = 299,792,458$ m

With Earth's Moon's observed orbital velocity value as:

$V_{emoon} = 1.022$ km/s or $1,022$ m

Earth's Moon's orbital quantizer Dc_emoon value becomes:

Dc_emoon = c / V_{emoon}
                = $299,792,458$ m / $1,022$ m   →

Dc_emoon = $293,339$

It is worth noting that the orbital quantizer Dc_emoon value of Earth's Moon is exactly an integer. This peculiarity reinforces the quantum nature of this mechanism.

Therefore Earth's Moon's variable radius Remoon becomes:

Remoon = [GMearth/c^2] x Dc_emoon^2
= 0.0044351999 m x 293,339^2 →

Remoon = 381,639,056 m or 381,639 km

which tightly matches Earth's Moon's observed orbital radius value as: 384,000 km (Δ=0.615%)

Orbital Velocity-Radius Verification Case of Earth's LAGEOS II:

LAGEOS-II is an Earth based satellite.

With LAGEOS-II's observed orbital velocity value as:

Vlageos2 = 5.70 km/s or 5,700 m

LAGEOS-II's orbital quantizer Dc_lageos2 value becomes:

Dc_lageos2 = c / Vlageos2
= 299,792,458 m / 5,700 m →

Dc_lageos2 = 52,595.16

Therefore LAGEOS-II's variable orbital radius Rlageos2 becomes:

Rlageos2 = [GMearth/c^2] x Dc_lageos2^2
= 0.0044351999 m x 52,595.16^2 →

Rlageos2 = 12,268,875 m or 12,268 km

which tightly matches LAGEOS-II's observed orbital radius value as: 12,270 km (Δ=0.017%) or 12,286 km (Δ=0.147%)

LAGEOS-II's orbital radius of 12,286 km is based on:
- Earth's radius of 6,378 km
- Periapsis altitude of 5,858 km
- Apoapsis altitude of 5,958 km
- 12,286 km = 6378 km + (5,858 km + 5,958 km)/2

Evidence Of Orbital Velocity-Radius Quantizer Mechanism Through Mars' Moons

Targets Of This Orbital Velocity-Radius Quantizer Mechanism Verification Of Mars' Moons:

Based on my calculations, there appears to exist a host of evidence of this orbital velocity-radius quantizer mechanism for all the known natural moons inside Mars' planetary system. The moons subjected to this verification are:

Phobos, Deimos

Reference Data Of This Orbital Velocity-Radius Verification of Mars' Moons:

The reference data here is:

Mars' Gravitational Energy Value $1GM/c^2$

Mars' effective gravitational energy potential quanta dedicated exclusively to orbital motion of each captive (E_{g1}) as:

GM_{mars}/c^2 = 0.0004766 m (4.7664809687338e-4 m)

based on values of:

M_{mars} = 6.4185e+23 kg (as mass of Mars)
G = 6.6743e-11 (N kg^{-2} m^2)
c = 299,792,458 m

Orbital Velocity-Radius Verification Case of Mars' Moon Phobos:

With Mars moon Phobos' observed orbital velocity value as:

V_{phobos} = 2.13849 km/s or 2,138.49 m

based on values of:

Phobos' semi-axis = 9,378 km
(→ circumference = 58,923.71 km)
Phobos' sidereal orbital period = 0.31891 days

(or 27,553.824 seconds)

Mars moon Phobos' orbital quantizer D_{c_phobos} value becomes:

D_{c_phobos}	$= c / V_{phobos}$
	$= 299{,}792{,}458 \text{ m} / 2{,}138.49 \text{ m}$ →
D_{c_phobos}	$= 140{,}189$ (exactly $140{,}188.85$)

Therefore Mars moon Phobos' variable radius R_{phobos} becomes:

R_{phobos}	$= [GM_{mars}/c^2] \times D_{c\_phobos}^2$
	$= 0.0004766 \text{ m} \times 140{,}189^2$ →
R_{phobos}	$= 9{,}366{,}598 \text{ m}$ or $9{,}366 \text{ km}$

which tightly matches Mars moon Phobos' observed orbital radius value as: $9{,}378 \text{ km}$ ($\Delta = 0.128\%$)

Orbital Velocity-Radius Verification Case of Mars Moon Deimos:

With Mars moon Deimos' observed orbital velocity value as:

$V_{deimos} = 1.35134 \text{ km/s}$ or $1{,}351.34 \text{ m}$

based on values of:

Deimos' semi-axis = $23{,}459 \text{ km}$
(→ circumference = $147{,}397.24 \text{ km}$)
Deimos' sidereal orbital period = 1.26244 days
(or $109{,}074.816$ seconds)

Mars moon Deimos' orbital quantizer D_{c_deimos} value becomes:

D_{c_deimos}	$= c / V_{deimos}$
	$= 299{,}792{,}458 \text{ m} / 1{,}351.34 \text{ m}$ →
D_{c_deimos}	$= 221{,}848$ (exactly $221{,}848.28$)

Therefore Mars moon Deimos' variable radius R_{deimos} becomes:

R_{deimos}	$= [GM_{mars}/c^2] \times D_{c\_deimos}^2$
	$= 0.0004766 \text{ m} \times 221{,}848^2$ →
R_{deimos}	$= 23{,}456{,}600 \text{ m}$ or $23{,}456 \text{ km}$

which tightly matches Mars moon Deimos' observed orbital radius value as: $23{,}459 \text{ km}$ ($\Delta = 0.012\%$)

Evidence Of Orbital Velocity-Radius Quantizer Mechanism Through Jupiter's Moons

Targets Of This Orbital Velocity-Radius Quantizer Mechanism Verification Of Jupiter's Moons:

Based on my calculations, there appears to exist a host of evidence of this orbital velocity-radius quantizer mechanism for the most known natural moons inside Jupiter's planetary system. The moons subjected to this verification are:

Metis, Adrastea, Amalthea, Thebe, Io, Europa, Ganymede, Callisto, Themisto, Leda, Himalia, Ersa, Pandia, Lysithea, Elara, DiaCarpo, Valetudo, Callirrhoe, Kallichore, S2003J9, Cyllene

Reference Data Of This Orbital Velocity-Radius Verification of Jupiter's Moons:

The reference data here is:

Jupiter's Gravitational Energy Value $1GM/c^2$

Jupiter's effective gravitational energy potential quanta dedicated exclusively to orbital motion of each captive (E_{g1}) as:

$GM_{jupiter}/c^2$ = 1.40993 m (exactly 1.40993078869 m)

based on values of:

$M_{jupiter}$ = 1.8986e+27 kg (as mass of Jupiter)
G = 6.6743e-11 (N kg^{-2} m²)
c = 299,792,458 m

Orbital Velocity-Radius Verification Case of Jupiter's Moon Metis:

With Metis' observed orbital velocity value as:

V_{metis} = 31.57763 km/s or 31,577.63 m

based on values of:

Metis' semi-axis = 128,000 km

($\rightarrow$ circumference = 804,247.72 km)
Metis' sidereal orbital period = 0.294779 days
(or 25,468.90 seconds)

Metis' orbital quantizer Dc_metis value becomes:

Dc_metis	= c / V_{metis}
	= 299,792,458 m / 31,577.63 m $\rightarrow$
Dc_metis	= 9494 (exactly 9493.82)

Therefore Metis' variable radius R_{metis} becomes:

R_{metis}	= [$GM_{jupiter}/c^2$] x Dc_metis^2
	= 1.40993 m x 9494^2 $\rightarrow$
R_{metis}	= 127,085,501 m or 127,085 km

which tightly matches Metis' observed orbital radius value as: 128,000 km km (Δ=0.719%)

Orbital Velocity-Radius Verification Case of Jupiter's Moon Adrastea:

With Adrastea's observed orbital velocity value as:

$V_{adrastea}$ = 31.452909 km/s or 31,452.91 m

based on values of:

Adrastea's semi-axis = 129,000 km
($\rightarrow$ circumference = 810,530.90 km)
Adrastea's sidereal orbital period = 0.298260 days
(or 25,769.664 seconds)

Adrastea's orbital quantizer $Dc\_adrastea$ value becomes:

$Dc\_adrastea$	= c / $V_{adrastea}$
	= 299,792,458 m / 31,452.91 m $\rightarrow$
$Dc\_adrastea$	= 9,531 (exactly 9,531.469)

Therefore Adrastea's variable radius $R_{adrastea}$ becomes:

$R_{adrastea}$	= [$GM_{jupiter}/c^2$] x $Dc\_adrastea^2$
	= 1.40993 m x $9,531^2$ $\rightarrow$
$R_{adrastea}$	= 128,077,986 m or 128,077 km

which tightly matches Adrastea's observed orbital radius value as: 129,000 km (Δ=0.10%)

Orbital Velocity-Radius Verification Case of Jupiter's Moon Amalthea:

With Amalthea's observed orbital velocity value as:

$V_{amalthea}$ = 26.48000 km/s or 26,480.00 m

based on values of:

Amalthea's semi-axis = 181,400 km
($\rightarrow$ circumference = 1,139,769.81 km)
Amalthea's sidereal orbital period = 0.498179 days
(or 43,042.665 seconds)

Amalthea's orbital quantizer $Dc\_amalthea$ value becomes:

$Dc\_amalthea$ = c / $V_{amalthea}$
= 299,792,458 m / 26,480.00 m $\rightarrow$

$Dc\_amalthea$ = 11,321 (exactly 11,321.46)

Therefore Amalthea's variable radius $R_{amalthea}$ becomes:

$R_{amalthea}$ = [$GM_{jupiter}/c^2$] x $Dc\_amalthea^2$
= 1.40993 m x $11,321^2$ $\rightarrow$

$R_{amalthea}$ = 180,703,736 m or 180,703 km

which tightly matches Amalthea's observed orbital radius value as: 181,400 km (Δ=0.38%)

Orbital Velocity-Radius Verification Case of Jupiter's Moon Thebe:

With Thebe's observed orbital velocity value as:

V_{thebe} = 23.92442 km/s or 23,924.42 m

based on values of:

Thebe's semi-axis = 221,900 km
($\rightarrow$ circumference = 1,394,238.81 km)
Thebe's sidereal orbital period = 0.6745 days
(or 58,276.8 seconds)

Thebe's orbital quantizer Dc_thebe value becomes:

Dc_thebe = c / V_{thebe}
= 299,792,458 m / 23,924.42 m $\rightarrow$

Dc_thebe = 12,531 (exactly 12,530.81)

Therefore Thebe's variable radius Rthebe becomes:

$$R_{thebe} = [GM_{jupiter}/c^2] \times D_{c\_thebe}^2 = 1.40993 \text{ m} \times 12{,}531^2 \rightarrow$$

$$R_{thebe} = 221{,}395{,}613 \text{ m or } 221{,}395 \text{ km}$$

which tightly matches Thebe's observed orbital radius value as: 221,900 km (Δ=0.23%)

Orbital Velocity-Radius Verification Case of Jupiter's Moon Io:

With Io's observed orbital velocity value as:

$$V_{io} = 17.33847 \text{ km/s or } 17{,}338.47 \text{ m}$$

based on values of:

Io's semi-axis = 421,800 km
(→ circumference = 2,650,247.56 km)
Io's sidereal orbital period = 1.769138 days
(or 152,853.52 seconds)

Io's orbital quantizer D_{c_io} value becomes:

$$D_{c\_io} = c / V_{io} = 299{,}792{,}458 \text{ m} / 17{,}338.47 \text{ m} \rightarrow$$

$$D_{c\_io} = 17{,}290 \text{ (exactly 17,290.59)}$$

Therefore Io's variable radius R_{io} becomes:

$$R_{io} = [GM_{jupiter}/c^2] \times D_{c\_io}^2 = 1.40993 \text{ m} \times 17{,}290^2 \rightarrow$$

$$R_{io} = 421{,}490{,}255 \text{ m or } 421{,}490 \text{ km}$$

which tightly matches Io's observed orbital radius value as: 421,800 km (Δ=0.073%)

Orbital Velocity-Radius Verification Case of Jupiter's Moon Europa:

With Europa's observed orbital velocity value as:

$$V_{europa} = 13.74296 \text{ km/s or } 13{,}742.96 \text{ m}$$

based on values of:

Europa's semi-axis = 671,100 km
(→ circumference = 4,216,645.66 km)

Europa's sidereal orbital period = 3.551181 days
(or 306,822.04 seconds)

Europa's orbital quantizer Dc_europa value becomes:

Dc_europa = c / Veuropa
= 299,792,458 m / 13,742.96 m →

Dc_europa = 21,814 (exactly 21,814.25)

Therefore Europa's variable radius Reuropa becomes:

Reuropa = $[GMjupiter/c^2]$ x Dc_europa^2
= 1.40993 m x $21,814^2$ →

Reuropa = 670,916,030 m or 670,916 km

which tightly matches Europa's observed orbital radius value as: 671,100 km (Δ=0.10%)

Orbital Velocity-Radius Verification Case of Jupiter's Moon Ganymede:

With Ganymede's observed orbital velocity value as:

Vganymede = 10.88002 km/s or 10,880.02 m

based on values of:

Ganymede's semi-axis = 1,070,400 km
(→ circumference = 6,725,521.55 km)
Ganymede's sidereal orbital period = 7.154553 days
(or 618,153.38 seconds)

Ganymede's orbital quantizer $Dc\_ganymede$ value becomes:

$Dc\_ganymede$ = c / Vganymede
= 299,792,458 m / 10,880.02 m →

$Dc\_ganymede$ = 27,554 (exactly 27,554.40)

Therefore Ganymede's variable radius Rganymede becomes:

Rganymede = $[GMjupiter/c^2]$ x $Dc\_ganymede^2$
= 1.40993 m x $27,554^2$ →

Rganymede = 1,070,451,166 m or 1,070,451 km

which tightly matches Ganymede's observed orbital radius value as: 1,070,400 km (Δ=0.004%)

Orbital Velocity-Radius Verification Case of Jupiter's Moon Callisto:

With Callisto's observed orbital velocity value as:

Vcallisto = 8.203826 km/s or 8,203.826 m

> based on values of:
>
> Callisto's semi-axis = 1,882,700 km
> ($\rightarrow$ circumference = 11,829,352.97 km)
> Callisto's sidereal orbital period = 16.689017 days
> (or 1,441,931.07 seconds)

Callisto's orbital quantizer Dc_callisto value becomes:

Dc_callisto	= c / Vcallisto
	= 299,792,458 m / 8,203.82 m $\rightarrow$
Dc_callisto	= 36,543 (exactly 36,543.03)

Therefore Callisto's variable radius Rcallisto becomes:

Rcallisto	= [GMjupiter/c^2] x Dc_callisto^2
	= 1.40993 m x 36,543^2 $\rightarrow$
Rcallisto	= 1,882,807,619 m or 1,882,807 km

> which tightly matches Callisto's observed orbital radius value as: 1,882,700 km (Δ=0.005%)

Orbital Velocity-Radius Verification Case of Jupiter's Moon Themisto:

With Themisto's observed orbital velocity value as:

Vthemisto = 4.19877 km/s or 4,198.77 m

> based on values of:
>
> Themisto's semi-axis = 7,507,000 km
> ($\rightarrow$ circumference = 47,167,872.10 km)
> Themisto's sidereal orbital period = 130.02 days
> (or 11,233,728 seconds)

Themisto's orbital quantizer Dc_themisto value becomes:

Dc_themisto	= c / Vthemisto
	= 299,792,458 m / 4,198.77 m $\rightarrow$
Dc_themisto	= 71,400 (exactly 71,400.06)

Therefore Themisto's variable radius Rthemisto becomes:

$$R_{themisto} = [GM_{jupiter}/c^2] \times Dc\_themisto^2$$
$$= 1.40993 \text{ m} \times 71,400^2 \quad \rightarrow$$

$$R_{themisto} = 7,187,766,742 \text{ m or } 7,187,766 \text{ km}$$

which tightly matches Themisto's observed orbital radius value as: 7,507,000 km (Δ=4.44%)

Orbital Velocity-Radius Verification Case of Jupiter's Moon Leda:

With Leda's observed orbital velocity value as:

$$V_{leda} = 3.37017 \text{ km/s or } 3,370.17 \text{ m}$$

based on values of:

Leda's semi-axis = 11,165,000 km
($\rightarrow$ circumference = 70,151,763.95 km)
Leda's sidereal orbital period = 240.92 days
(or 20,815,488 seconds)

Leda's orbital quantizer Dc_leda value becomes:

$$Dc\_leda = c / V_{leda}$$
$$= 299,792,458 \text{ m} / 3,370.17 \text{ m} \quad \rightarrow$$

$$Dc\_leda = 88,954 \text{ (exactly } 88,954.69)$$

Therefore Leda's variable radius R_{leda} becomes:

$$R_{leda} = [GM_{jupiter}/c^2] \times Dc\_leda^2$$
$$= 1.40993 \text{ m} \times 88,954^2 \quad \rightarrow$$

$$R_{leda} = 11,156,514,006 \text{ m or } 11,156,514 \text{ km}$$

which tightly matches Leda's observed orbital radius value as: 11,165,000 km (Δ=0.076%)

Orbital Velocity-Radius Verification Case of Jupiter's Moon Himalia:

With Himalia's observed orbital velocity value as:

$$V_{himalia} = 3.32633 \text{ km/s or } 3,326.33 \text{ m}$$

based on values of:

Himalia's semi-axis = 11,461,000 km
($\rightarrow$ circumference = 72,011,586.80 km)
Himalia's sidereal orbital period = 250.5662 days

(or 21,648,919 seconds)

Himalia's orbital quantizer $D_{c\_himalia}$ value becomes:

$D_{c\_himalia}$ = c / $V_{himalia}$
= 299,792,458 m / 3,326.33 m →

$D_{c\_himalia}$ = 90,126 (exactly 90,126.92)

Therefore Himalia's variable radius $R_{himalia}$ becomes:

$R_{himalia}$ = [GMjupiter/c^2] x $D_{c\_himalia}^2$
= 1.40993 m x 90,126^2 →

$R_{himalia}$ = 11,452,432,596 m or 11,452,432 km

which tightly matches Himalia's observed orbital radius value as: 11,461,000 km (Δ=0.074%)

Orbital Velocity-Radius Verification Case of Jupiter's Moon Ersa:

With Ersa's observed orbital velocity value as:

V_{ersa} = 3.31418 km/s or 3,314.18 m

based on values of:

Ersa's semi-axis = 11,483,000 km
(→ circumference = 72,149,816.88 km)
Ersa's sidereal orbital period = 252.0 days
(or 21,770,000 seconds)

Ersa's orbital quantizer D_{c_ersa} value becomes:

D_{c_ersa} = c / V_{ersa}
= 299,792,458 m / 3,314.18 m →

D_{c_ersa} = 90,457 (exactly 90,457.50)

Therefore Ersa's variable radius R_{ersa} becomes:

R_{ersa} = [GMjupiter/c^2] x $D_{c\_ersa}^2$
= 1.40993 m x 90,457^2 →

R_{ersa} = 11,536,708,304 m or 11,536,708 km

which tightly matches Ersa's observed orbital radius value as: 11,483,000 km (Δ=0.467%)

Orbital Velocity-Radius Verification Case of Jupiter's Moon Pandia:

With Pandia's observed orbital velocity value as:

Vpandia = 3.32456 km/s or 3,324.56 m

> based on values of:

Pandia's semi-axis = 11,525,000 km
($\rightarrow$ circumference = 72,413,710.66 km)
Pandia's sidereal orbital period = 252.1 days
(or 21,781,440 seconds)

Pandia's orbital quantizer Dc_pandia value becomes:

Dc_pandia   = c / Vpandia
                 = 299,792,458 m / 3,324.56 m $\rightarrow$

Dc_pandia   = 90,175 (exactly 90,175.07)

Therefore Pandia's variable radius Rpandia becomes:

Rpandia   = [GMjupiter/c^2] x Dc_pandia^2
               = 1.40993 m x 90,175^2 $\rightarrow$

Rpandia   = 11,464,888,974 m or 11,464,888 km

> which tightly matches Pandia's observed orbital radius value as: 11,525,000 km (Δ=0.524%)

Orbital Velocity-Radius Verification Case of Jupiter's Moon Lysithea:

With Lysithea's observed orbital velocity value as:

Vlysithea = 3.28710 km/s or 3,287.10 m

> based on values of:

Lysithea's semi-axis = 11,717,000 km
($\rightarrow$ circumference = 73,620,082.24 km)
Lysithea's sidereal orbital period = 259.22 days
(or 22,396,608 seconds)

Lysithea's orbital quantizer Dc_lysithea value becomes:

Dc_lysithea   = c / Vlysithea
                   = 299,792,458 m / 3,287.10 m $\rightarrow$

Dc_lysithea   = 91,202 (exactly 91,202.48)

Therefore Lysithea's variable radius Rlysithea becomes:

Rlysithea   = [GMjupiter/c^2] x Dc_lysithea^2

$$= 1.40993 \text{ m} \times 91{,}202^2 \quad \rightarrow$$

R$_{lysithea}$ $\quad = 11{,}727{,}522{,}527$ m or $11{,}727{,}522$ km

which tightly matches Lysithea's observed orbital radius value as: $11{,}717{,}000$ km ($\Delta=0.089\%$)

Orbital Velocity-Radius Verification Case of Jupiter's Moon Elara:

With Elara's observed orbital velocity value as:

V$_{elara}$ = 3.28835 km/s or 3,288.35 m

based on values of:

Elara's semi-axis = 11,741,000 km
($\rightarrow$ circumference = 73,770,878.69 km)
Elara's sidereal orbital period = 259.6528 days
(or 22,434,002 seconds)

Elara's orbital quantizer Dc_elara value becomes:

Dc_elara	= c / V$_{elara}$
	= 299,792,458 m / 3,288.35 m $\quad \rightarrow$
Dc_elara	= 91,168 (exactly 91,168.01)

Therefore Elara's variable radius R$_{elara}$ becomes:

R$_{elara}$	= [GMjupiter/c^2] x Dc_elara^2
	= 1.40993 m x 91,168^2 $\quad \rightarrow$
R$_{elara}$	= 11,718,780,143 m or 11,718,780 km

which tightly matches Elara's observed orbital radius value as: $11{,}741{,}000$ km ($\Delta=0.189\%$)

Orbital Velocity-Radius Verification Case of Jupiter's Moon Dia:

With Dia's observed orbital velocity value as:

V$_{dia}$ = 3.07054 km/s or 3,070.54 m

based on values of:

Dia's semi-axis = 12,118,000 km
($\rightarrow$ circumference = 76,139,639.55 km)
Dia's sidereal orbital period = 287.0 days
(or 24,796,800 seconds)

Dia's orbital quantizer D_{c_dia} value becomes:

$$\begin{aligned}D_{c\_dia} &= c / V_{dia} \\ &= 299{,}792{,}458 \text{ m} / 3{,}070.14 \text{ m} \quad \rightarrow \\ D_{c\_dia} &= 97{,}635 \text{ (exactly } 97{,}635.09)\end{aligned}$$

Therefore Dia's variable radius R_{dia} becomes:

$$\begin{aligned}R_{dia} &= [GM_{jupiter}/c^2] \times D_{c\_dia}^2 \\ &= 1.40993 \text{ m} \times 97{,}635^2 \quad \rightarrow \\ R_{dia} &= 13{,}440{,}289{,}165 \text{ m or } 13{,}440{,}289 \text{ km}\end{aligned}$$

which tightly matches Dia's observed orbital radius value as: 12,118,000 km (Δ=10.91%)

Orbital Velocity-Radius Verification Case of Jupiter's Moon Carpo:

With Carpo's observed orbital velocity value as:

V_{carpo} = 2.70878 km/s or 2,708.78 m

based on values of:

Carpo's semi-axis = 16,989,000 km
(→ circumference = 106,745,035.18 km)
Carpo's sidereal orbital period = 456.1 days
(or 39,407,040 seconds)

Carpo's orbital quantizer D_{c_carpo} value becomes:

$$\begin{aligned}D_{c\_carpo} &= c / V_{carpo} \\ &= 299{,}792{,}458 \text{ m} / 2{,}708.78 \text{ m} \quad \rightarrow \\ D_{c\_carpo} &= 110{,}674 \text{ (exactly } 110{,}674.31)\end{aligned}$$

Therefore Carpo's variable radius R_{carpo} becomes:

$$\begin{aligned}R_{carpo} &= [GM_{jupiter}/c^2] \times D_{c\_carpo}^2 \\ &= 1.40993 \text{ m} \times 110{,}674^2 \quad \rightarrow \\ R_{carpo} &= 17{,}269{,}857{,}917 \text{ m or } 17{,}269{,}857 \text{ km}\end{aligned}$$

which tightly matches Carpo's observed orbital radius value as: 16,989,000 km (Δ=1.653%)

Orbital Velocity-Radius Verification Case of Jupiter's Moon Valetudo:

With Valetudo's observed orbital velocity value as:

$V_{valetudo}$ = 2.58815 km/s or 2,588.15 m

based on values of:

Valetudo's semi-axis = 18,980,000 km
(→ circumference = 119,254,857.13 km)
Valetudo's sidereal orbital period = 533.3 days
(or 46,077,120 seconds)

Valetudo's orbital quantizer $Dc\_valetudo$ value becomes:

$Dc\_valetudo$ = c / $V_{valetudo}$
= 299,792,458 m / 2,588.15 m →

$Dc\_valetudo$ = 115,832 (exactly 115,832.37)

Therefore Valetudo's variable radius $R_{valetudo}$ becomes:

$R_{valetudo}$ = $[GM_{jupiter}/c^2]$ x $Dc\_valetudo^2$
= 1.40993 m x $115,832^2$ →

$R_{valetudo}$ = 18,917,104,442 m or 18,917,104 km

which tightly matches Valetudo's observed orbital radius value as: 18,980,000 km (Δ=0.33%)

Orbital Velocity-Radius Verification Case of Jupiter's Moon Callirrhoe:

With Callirrhoe's observed orbital velocity value as:

$V_{callirrhoe}$ = 2.30989 km/s or 2,309.89 m

based on values of:

Callirrhoe's semi-axis = 24,102,000 km
(→ circumference = 151,437,332.27 km)
Callirrhoe's sidereal orbital period = 758.8 days
(or 65,560,320 seconds)

Callirrhoe's orbital quantizer $Dc\_callirrhoe$ value becomes:

$Dc\_callirrhoe$ = c / $V_{callirrhoe}$
= 299,792,458 m / 2,309.89 m →

$Dc\_callirrhoe$ = 129,786 (exactly 129,786.46)

Therefore Callirrhoe's variable radius $R_{callirrhoe}$ becomes:

$R_{callirrhoe}$ = $[GM_{jupiter}/c^2]$ x $Dc\_callirrhoe^2$

$$= 1.40993 \text{ m} \times 129{,}786^2 \quad \rightarrow$$

Rcallirrhoe $= 23{,}749{,}433{,}063$ m or $23{,}749{,}433$ km

which tightly matches Callirrhoe's observed orbital radius value as: $24{,}102{,}000$ km ($\Delta = 1.48\%$)

Important Notice:
It appears that the observed orbital period and radius of Kallichore and the ones of Callirrhoe are inconsistent.
This is because Callirrhoe's orbital period is shorter than Kallichore's one, and yet Callirrhoe's orbital radius is longer than Kallichore's one. Either one of them is wrong.

Orbital Velocity-Radius Verification Case of Jupiter's Moon Kallichore:

With Kallichore's observed orbital velocity value as:

Vkallichore $= 2.28646$ km/s or $2{,}286.46$ m

based on values of:

Kallichore's semi-axis $= 24{,}043{,}000$ km
($\rightarrow$ circumference $= 151{,}066{,}624.34$ km)
Kallichore's sidereal orbital period $= 764.7$ days
(or $66{,}070{,}080$ seconds)

Kallichore's orbital quantizer Dc_kallichore value becomes:

Dc_kallichore $= c$ / Vkallichore
$= 299{,}792{,}458$ m / $2{,}286.46$ m $\quad \rightarrow$

Dc_kallichore $= 131{,}116$ (exactly $131{,}116.42$)

Therefore Kallichore's variable radius Rkallichore becomes:

Rkallichore $= [\text{GMjupiter}/c^2] \times \text{Dc\_kallichore}^2$
$= 1.40993$ m $\times 131{,}116^2 \quad \rightarrow$

Rkallichore $= 24{,}238{,}678{,}294$ m or $24{,}238{,}678$ km

which tightly matches Kallichore's observed orbital radius value as: $24{,}043{,}000$ km ($\Delta = 0.81\%$)

Important Notice:
It appears that the observed orbital period and radius of Kallichore and the ones of Callirrhoe are inconsistent.
This is because Callirrhoe's orbital period is shorter than Kallichore's one, and yet Callirrhoe's orbital radius is longer than Kallichore's one.

Either one of them is wrong.

Orbital Velocity-Radius Verification Case of Jupiter's Moon S2003J9:

With S2003J9's observed orbital velocity value as:

$V_{s2003j9}$ = 2.29921 km/s or 2,299.21 m

based on values of:

S2003J9's semi-axis = 24,234,000 km
(→ circumference = 152,266,712.73 km)
S2003J9's sidereal orbital period = 766.5 days
(or 66,225,600 seconds)

S2003J9's orbital quantizer $Dc\_s2003j9$ value becomes:

$Dc\_s2003j9$	= c / $V_{s2003j9}$
	= 299,792,458 m / 2,299.21 m →
$Dc\_s2003j9$	= 130,389 (exactly 130,389.33)

Therefore S2003J9's variable radius $Rs2003j9$ becomes:

$Rs2003j9$	= [$GM_{jupiter}/c^2$] x $Dc\_s2003j9^2$
	= 1.40993 m x 130,389^2 →
$Rs2003j9$	= 23,970,630,672 m or 23,970,630 km

which tightly matches S2003J9's observed orbital radius value as: 24,234,000 km (Δ=1.09%)

Orbital Velocity-Radius Verification Case of Jupiter's Moon Cyllene:

With Cyllene's observed orbital velocity value as:

$V_{cyllene}$ = 2.39998 km/s or 2,399.98 m

based on values of:

Cyllene's semi-axis = 24,349,000 km
(→ circumference = 152,989,279.04 km)
Cyllene's sidereal orbital period = 737.8 days
(or 63,745,920 seconds)

Cyllene's orbital quantizer $Dc\_cyllene$ value becomes:

| $Dc\_cyllene$ | = c / $V_{cyllene}$ |
| | = 299,792,458 m / 2,399.98 m →|

Dc_cyllene = 124,914 (exactly 124,914.56)

Therefore Cyllene's variable radius Rcyllene becomes:

Rcyllene = [GMjupiter/c^2] x Dc_cyllene^2
= 1.40993 m x 124,914^2 →

Rcyllene = 21,999,853,182 m or 21,999,853 km

which tightly matches Cyllene's observed orbital radius value as: 24,349,000 km (Δ=10.67%)

Evidence Of Orbital Velocity-Radius Quantizer Mechanism Through Saturn's Moons

Targets Of This Orbital Velocity-Radius Quantizer Mechanism Verification Of Saturn's Moons:

Based on my calculations, there appears to exist a host of evidence of this orbital velocity-radius quantizer mechanism for the most known natural moons inside Saturn's planetary system. The moons subjected to this verification are:

    Mimas, Enceladus, Tethys, Dione, Rhea, Titan, Hyperion, Iapetus

Reference Data Of This Orbital Velocity-Radius Verification of Saturn's Moons:

The reference data here is:

Saturn's Gravitational Energy Value $1GM/c^2$

Saturn's effective gravitational energy potential quanta dedicated exclusively to orbital motion of each captive (E_{g1}) as:

    GM_{saturn}/c^2 = 0.422147 m (exactly 0.4221475066 m)

    based on values of:

    M_{saturn} = 5.6846e+26 kg (as mass of Saturn)
    G = 6.6743e-11 (N kg^{-2} m^2)
    c = 299,792,458 m

Orbital Velocity-Radius Verification Case of Saturn's Moon Mimas:

    With Mimas' observed orbital velocity value as:

    V_{mimas} = 14.31566 km/s or 14,315.66 m

    based on values of:

    Mimas' semi-axis = 185,520 km
    ($\rightarrow$ circumference = 1,165,656.53 km)
    Mimas' sidereal orbital period = 0.94242180 days

(or 81,425.24352 seconds)

Mimas' orbital quantizer Dc_mimas value becomes:

Dc_mimas = c / V_{mimas}
= 299,792,458 m / 14,315.66 m →

Dc_mimas = 20,941 (exactly 20,941.57)

Therefore Mimas' variable radius R_{mimas} becomes:

R_{mimas} = [GM_{saturn}/c^2] x Dc_mimas^2
= 0.422147 m x $20,941^2$ →

R_{mimas} = 185,122,216 m or 185,122 km

which tightly matches Mimas' observed orbital radius value as: 185,520 km (Δ=0.214%)

Orbital Velocity-Radius Verification Case of Saturn's Moon Enceladus:

With Enceladus' observed orbital velocity value as:

$V_{enceladus}$ = 12.63251 km/s or 12,632.51 m

based on values of:

Enceladus' semi-axis = 238,020 km
(→ circumference = 1,495,523.76 km)
Enceladus' sidereal orbital period = 1.370218 days
(or 118,386.8352 seconds)

Enceladus' orbital quantizer $Dc\_enceladus$ value becomes:

$Dc\_enceladus$ = c / $V_{enceladus}$
= 299,792,458 m / 12,632.51 m →

$Dc\_enceladus$ = 23,731 (exactly 23,731.82)

Therefore Enceladus' variable radius $R_{enceladus}$ becomes:

$R_{enceladus}$ = [GM_{saturn}/c^2] x $Dc\_enceladus^2$
= 0.422147 m x $23,731^2$ →

$R_{enceladus}$ = 237,736,456 m or 237,736 km

which tightly matches Enceladus' observed orbital radius value as: 238,020 km (Δ=0.119%)

Orbital Velocity-Radius Verification Case of Saturn's Moon Tethys:

With Tethys' observed orbital velocity value as:

Vtethys = 11.35091 km/s or 11,350.91 m

based on values of:

Tethys' semi-axis = 294,660 km
(→ circumference = 1,851,403.38 km)
Tethys' sidereal orbital period = 1.887802 days
(or 163,106.0928 seconds)

Tethys' orbital quantizer Dc_tethys value becomes:

Dc_tethys	= c / Vtethys
	= 299,792,458 m / 11,350.91 m →
Dc_tethys	= 26,411 (exactly 26,411.30)

Therefore Tethys' variable radius Rtethys becomes:

Rtethys	= [GMsaturn/c^2] x Dc_tethys^2
	= 0.422147 m x 26,411^2 →
Rtethys	= 294,464,807 m or 294,464 km

which tightly matches Tethys' observed orbital radius value as: 294,660 km (Δ=0.066%)

Orbital Velocity-Radius Verification Case of Saturn's Moon Dione:

With Dione's observed orbital velocity value as:

Vdione = 10.02782 km/s or 10,027.82 m

based on values of:

Dione's semi-axis = 377,400 km
(→ circumference = 2,371,274.13 km)
Dione's sidereal orbital period = 2.736915 days
(or 236,469.456 seconds)

Dione's orbital quantizer Dc_dione value becomes:

Dc_dione	= c / Vdione
	= 299,792,458 m / 10,027.82 m →
Dc_dione	= 29,896 (exactly 29,896.06)

Therefore Dione's variable radius Rdione becomes:

$$R_{dione} = [GM_{saturn}/c^2] \times D_{c\_dione}^2$$
$$= 0.422147 \text{ m} \times 29,896^2 \quad \rightarrow$$

R_{dione} = 377,302,668 m or 377,302 km

which tightly matches Dione's observed orbital radius value as: 377,400 km (Δ=0.025%)

Orbital Velocity-Radius Verification Case of Saturn's Moon Rhea:

With Rhea's observed orbital velocity value as:

V_{rhea} = 8.48421 km/s or 8,484.21 m

based on values of:

Rhea's semi-axis = 527,040 km
($\rightarrow$ circumference = 3,311,489.98 km)
Rhea's sidereal orbital period = 4.517500 days
(or 390,312 seconds)

Rhea's orbital quantizer $D_{c\,rhea}$ value becomes:

$$D_{c\_rhea} = c / V_{rhea}$$
$$= 299,792,458 \text{ m} / 8,484.21 \text{ m} \quad \rightarrow$$

D_{c_rhea} = 35,335 (exactly 35,335.34)

Therefore Rhea's variable radius R_{rhea} becomes:

$$R_{rhea} = [GM_{saturn}/c^2] \times D_c^{\;2}$$
$$= 0.422147 \text{ m} \times 35,335^2 \quad \rightarrow$$

R_{rhea} = 527,076,797 m or 527,076 km

which tightly matches Rhea's observed orbital radius value as: 527,040 km (Δ=0.0068%)

Orbital Velocity-Radius Verification Case of Saturn's Moon Titan:

With Titan's observed orbital velocity value as:

V_{titan} = 5.57256 km/s or 5,572.56 m

based on values of:

Titan's semi-axis = 1,221,870 km
($\rightarrow$ circumference = 7,677,235.63 km)
Titan's sidereal orbital period = 15.945421 days

(or 1,377,684.3744 seconds)

Titan's orbital quantizer D_{c_titan} value becomes:

$$D_{c\_titan} = c / V_{titan}$$
$$= 299{,}792{,}458 \text{ m} / 5{,}572.56 \text{ m} \quad \rightarrow$$

$$D_{c\_titan} = 53{,}797 \text{ (exactly 53,797.97)}$$

Therefore Titan's variable radius R_{titan} becomes:

$$R_{titan} = [GM_{saturn}/c^2] \times D_{c\_titan}^2$$
$$= 0.422147 \text{ m} \times 53{,}797^2 \quad \rightarrow$$

$$R_{titan} = 1{,}221{,}742{,}897 \text{ m or } 1{,}221{,}742 \text{ km}$$

which tightly matches Titan's observed orbital radius value as: 1,221,870 km (Δ=0.010%)

Orbital Velocity-Radius Verification Case of Saturn's Moon Hyperion:

With Hyperion's observed orbital velocity value as:

$V_{hyperion}$ = 5.13008 km/s or 5,130.08 m

based on values of:

Hyperion's semi-axis = 1,500,930 km
($\rightarrow$ circumference = 9,430,621.32 km)
Hyperion's sidereal orbital period = 21.276609 days
(or 1,838,299.0176 seconds)

Hyperion's orbital quantizer $D_{c\_hyperion}$ value becomes:

$$D_{c\_hyperion} = c / V_{hyperion}$$
$$= 299{,}792{,}458 \text{ m} / 5{,}130.08 \text{ m} \quad \rightarrow$$

$$D_{c\_hyperion} = 58{,}438 \text{ (exactly 58,438.16)}$$

Therefore Hyperion's variable radius $R_{hyperion}$ becomes:

$$R_{hyperion} = [GM_{saturn}/c^2] \times D_{c\_hyperion}^2$$
$$= 0.422147 \text{ m} \times 58{,}438^2 \quad \rightarrow$$

$$R_{hyperion} = 1{,}441{,}631{,}939 \text{ m or } 1{,}441{,}631 \text{ km}$$

which tightly matches Hyperion's observed orbital radius value as: 1,500,930 km (Δ=04.113%)

Orbital Velocity-Radius Verification Case of Saturn's Moon Iapetus:

With Iapetus' observed orbital velocity value as:

$V_{iapetus}$ = 3.26423 km/s or 3,264.23 m

   based on values of:

   Iapetus' semi-axis = 3,560,850 km
   ($\rightarrow$ circumference = 22,373,480.40 km)
   Iapetus' sidereal orbital period = 79.330183 days
      (or 6,854,127.8112 seconds)

Iapetus' orbital quantizer $D_{c\_iapetus}$ value becomes:

$D_{c\_iapetus}$	= c / $V_{iapetus}$
	= 299,792,458 m / 3,264.23 m $\rightarrow$
$D_{c\_iapetus}$	= 91,841 (exactly 91,841.70)

Therefore Iapetus' variable radius $R_{iapetus}$ becomes:

$R_{iapetus}$	= [GM_{saturn}/c^2] x $D_{c\_iapetus}^2$
	= 0.422147 m x 91,841² $\rightarrow$
$R_{iapetus}$	= 3,560,712,547 m or 3,560,712 km

   which tightly matches Iapetus' observed orbital radius value as: 3,560,850 km (Δ=0.0038%)

Evidence Of Orbital Velocity-Radius Quantizer Mechanism Through Uranus' Moons

Targets Of This Orbital Velocity-Radius Quantizer Mechanism Verification Of Uranus' Moons:

Based on my calculations, there appears to exist a host of evidence of this orbital velocity-radius quantizer mechanism for the most known natural moons inside Uranus' planetary system. The moons subjected to this verification are:

Miranda, Ariel, Umbriel, Titania, Oberon

Reference Data Of This Orbital Velocity-Radius Verification of Uranus' Moons:

The reference data here is:

Uranus' Gravitational Energy Value $1GM/c^2$

Uranus' effective gravitational energy potential quanta dedicated exclusively to orbital motion of each captive (E_{g1}) as:

GM_{uranus}/c^2 = 0.064466 m (exactly 0.06446649729 m)

based on values of:

M_{uranus} = 8.6810e+25 kg (as mass of Uranus)
G = 6.6743e-11 (N kg^{-2} m²)
c = 299,792,458 m

Orbital Velocity-Radius Verification Case of Uranus' Moon Miranda:

With Miranda's observed orbital velocity value as:

$V_{miranda}$ = 6.6832 km/s or 6,683.22 m

based on values of:

Miranda's semi-axis = 129,900 km
(→ circumference = 816,185.77 km)
Miranda's sidereal orbital period = 1.413479 days

(or 122,124.5856 seconds)

Miranda's orbital quantizer $Dc\_miranda$ value becomes:

$Dc\_miranda$ $= c / V_{miranda}$
                      $= 299{,}792{,}458$ m $/ 6{,}683.22$ m $\rightarrow$

$Dc\_miranda$ $= 44{,}857$ (exactly $44{,}857.48$)

Therefore Miranda's variable radius $R_{miranda}$ becomes:

$R_{miranda}$ $= [GM_{uranus}/c^2] \times Dc\_miranda^2$
                 $= 0.064466$ m $\times 44{,}857^2$ $\rightarrow$

$R_{miranda}$ $= 129{,}715{,}290$ m or $129{,}715$ km

which tightly matches Miranda's observed orbital radius value as: $129{,}900$ km ($\Delta = 0.100\%$)

Orbital Velocity-Radius Verification Case of Uranus' Moon Ariel:

With Ariel's observed orbital velocity value as:

$V_{ariel} = 5.50815$ km/s or $5{,}508.15$ m

based on values of:

Ariel's semi-axis $= 190{,}900$ km
($\rightarrow$ circumference $= 1{,}199{,}460.07$ km)
Ariel's sidereal orbital period $= 2.520379$ days
(or $217{,}760.7456$ seconds)

Ariel's orbital quantizer Dc_ariel value becomes:

Dc_ariel $= c / V_{ariel}$
               $= 299{,}792{,}458$ m $/ 5{,}508.15$ m $\rightarrow$

Dc_ariel $= 54{,}427$ (exactly $54{,}427.06$)

Therefore Ariel's variable radius R_{ariel} becomes:

R_{ariel} $= [GM_{uranus}/c^2] \times Dc\_ariel^2$
             $= 0.064466$ m $\times 54{,}427^2$ $\rightarrow$

R_{ariel} $= 190{,}968{,}005$ m or $190{,}968$ km

which tightly matches Ariel's observed orbital radius value as: $190{,}900$ km ($\Delta = 0.035\%$)

Orbital Velocity-Radius Verification Case of Uranus' Moon Umbriel:

With Umbriel's observed orbital velocity value as:

Vumbriel = 4.66777 km/s or 4,667.77 m

    based on values of:

Umbriel's semi-axis = 266,000 km
    (→ circumference = 1,671,327.29 km)
Umbriel's sidereal orbital period = 4.144176 days
    (or 358,056.8064 seconds)

Umbriel's orbital quantizer Dc_umbriel value becomes:

Dc_umbriel	= c / Vumbriel
	= 299,792,458 m / 4,667.77 m →
Dc_umbriel	= 64,226 (exactly 64,226.03)

Therefore Umbriel's variable radius Rumbriel becomes:

Rumbriel	= [GMuranus/c^2] x Dc_umbriel^2
	= 0.064466 m x 64,226^2 →
Rumbriel	= 265,921,215 m or 265,921 km

    which tightly matches Umbriel's observed orbital radius
    value as: 266,000 km (Δ=0.029%)

Orbital Velocity-Radius Verification Case of Uranus' Moon Titania:

With Titania's observed orbital velocity value as:

Vtitania = 3.64451 km/s or 3,644.51 m

    based on values of:

Titania's semi-axis = 436,300 km
    (→ circumference = 2,741,353.75 km)
Titania's sidereal orbital period = 8.705867 days
    (or 752,186.9088 seconds)

Titania's orbital quantizer Dc_titania value becomes:

Dc_titania	= c / Vtitania
	= 299,792,458 m / 3,644.51 m →
Dc_titania	= 82,258 (exactly 82,258.61)

Therefore Titania's variable radius Rtitania becomes:

$R_{titania}$ = [GM_{uranus}/c^2] x $Dc\_titania^2$
= 0.064466 m x 82,258² →

$R_{titania}$ = 436,207,881 m or 436,207 km

which tightly matches Titania's observed orbital radius value as: 436,300 km (Δ=0.021%)

Orbital Velocity-Radius Verification Case of Uranus' Moon Oberon:

With Oberon's observed orbital velocity value as:

V_{oberon} = 3.15179 km/s or 3,151.79 m

based on values of:

Oberon's semi-axis = 583,500 km
(→ circumference = 3,666,238.62 km)
Oberon's sidereal orbital period = 13.463234 days
(or 1,163,223.4176 seconds)

Oberon's orbital quantizer Dc_oberon value becomes:

Dc_oberon = c / V_{oberon}
= 299,792,458 m / 3,151.79 m →

Dc_oberon = 95,118 (exactly 95,118.08)

Therefore Oberon's variable radius R_{oberon} becomes:

R_{oberon} = 95,118² x 0.064466 m, hence:

R_{oberon} = 583,252,952 m or 583,252 km

R_{oberon} = [GM_{uranus}/c^2] x Dc_oberon^2
= 0.064466 m x 95,118² →

R_{oberon} = 583,252,952 m or 583,252 km

which tightly matches Oberon's observed orbital radius value as: 583,500 km (Δ=0.042%)

Evidence Of Orbital Velocity-Radius Quantizer Mechanism Through TOI-178's Exoplanets

Targets Of This Orbital Velocity-Radius Quantizer Mechanism Verification Of TOI-178's Exoplanets:

Based on my calculations, there appears to exist a host of evidence of this orbital velocity-radius quantizer mechanism for the most known exoplanets inside TOI-178's star system. The exoplanets subjected to this verification are:

TOI-178 b, TOI-178 c, TOI-178 d, TOI-178 e, TOI-178 f, TOI-178 g

Reference Data Of This Orbital Velocity-Radius Verification of TOI-178's Exoplanets:

The reference data here is:

TOI-178's Gravitational Energy Value $1GM/c^2$

TOI-178's effective gravitational energy potential quanta dedicated exclusively to orbital motion of each captive (E_{g1}) as:

GM_{toi178}/c^2 = 959.80663634384929 m

based on values of:

M_{toi178} = 1.292525e+30 kg (as mass of TOI-178)
G = 6.6743e-11 (N kg^{-2} m²)
c = 299,792,458 m

Orbital Velocity-Radius Verification Case Of Planet TOI-178 b:

With TOI-178 b planet's observed orbital velocity value as:

$V_{toi178b}$ = 148.13927 km/s or 148,139.27 m

based on values of:

TOI-178 b planet's semi-axis = 3,900,072 km (or 0.02607 AU)
→ circumference = 24,504,875.08734 km)

TOI-178 b planet's orbital period = 1.914558 days

(or 165,417.8112 seconds)

TOI-178 b planet's orbital quantizer $D_{c\_toi178b}$ value becomes:

$D_{c\_toi178b}$ = c / $V_{toi178b}$
= 299,792,458 m / 148,139.27 m →

$D_{c\_toi178b}$ = 2,023 (exactly 2,023.72)

Therefore TOI-178 b planet's variable radius $R_{toi178b}$ becomes:

$R_{toi178b}$ = [GM_{toi178}/c^2] x $D_{c\_toi178b}^2$
= 959.80663634384929 m x $2,023^2$ →

$R_{toi178b}$ = 3,928,036,493 m or 3,928,036 km

which tightly matches TOI-178 b planet's observed orbital radius value as: 3,900,072 km (Δ=0.71%)

Orbital Velocity-Radius Verification Case Of Planet TOI-178 c:

With TOI-178 c Planet's observed orbital velocity value as:

$V_{toi178c}$ = 124.29745 km/s or 124,297.45 m

based on values of:

TOI-178 c Planet's semi-axis = 5,535,200 km (or 0.037 AU)
(→ circumference = 34,778,687.31 km)

TOI-178 c Planet's orbital period = 3.23845 days
(or 279,802.08 seconds)

TOI-178 c Planet's orbital quantizer $D_{c\_toi178c}$ value becomes:

$D_{c\_toi178c}$ = c / $V_{toi178c}$
= 299,792,458 m / 124,297.45 m →

$D_{c\_toi178c}$ = 2,411 (exactly 2,411.89)

Therefore TOI-178 c Planet's variable radius $R_{toi178c}$ becomes:

$R_{toi178c}$ = [GM_{toi178}/c^2] x $D_{c\_toi178c}^2$
= 959.80663634384929 m x $2,411^2$ →

$R_{toi178c}$ = 5,579,280,152 m or 5,579,280 km

which tightly matches TOI-178 c Planet's observed orbital radius value as: 5,535,200 km (Δ=0.796%)

Orbital Velocity-Radius Verification Case Of Planet TOI-178 d:

With TOI-178 d planet's observed orbital velocity value as:

$V_{toi178d}$ = 98.21275 km/s or 98,212.75 m

> based on values of:
>
>> TOI-178 d planet's semi-axis = 8,856,320 km (or 0.0592 AU)
>> ($\rightarrow$ circumference = 55,645,899.69 km)
>>
>> TOI-178 d planet's sidereal orbital period = 6.5577 days
>> (or 566,585.28 seconds)

TOI-178 d planet's orbital quantizer $D_{ctoi178d}$ value becomes:

$D_{c\_toi178d}$ = c / $V_{toi178d}$
= 299,792,458 m / 98,212.75 m $\rightarrow$

$D_{c\_toi178d}$ = 3,052 (exactly 3,052.48)

Therefore TOI-178 d planet's variable radius $R_{toi178d}$ becomes:

$R_{toi178d}$ = [GM_{toi178}/c^2] x $D_{c\_toi178d}^2$
= 959.80663634384929 m x $3,052^2$ $\rightarrow$

$R_{toi178d}$ = 8,940,314,714 m or 8,940,314 km

> which tightly matches TOI-178 d planet's observed orbital radius value as: 8,856,320 km (Δ=0.948%)

Orbital Velocity-Radius Verification Case Of Planet TOI-178 e:

> With TOI-178 e planet's observed orbital velocity value as:

$V_{toi178e}$ = 85.51024 km/s or 85,510.24 m

> based on values of:
>
>> TOI-178 e planet's semi-axis = 11,713,680 km (or 0.0783 AU)
>> ($\rightarrow$ circumference = 73,599,222.06 km)
>>
>> TOI-178 e planet's sidereal orbital period = 9.961881 days
>> (or 860,706.51 seconds)

TOI-178 e planet's orbital quantizer $D_{ctoi178e}$ value becomes:

$D_{c\_toi178e}$ = c / $V_{toi178e}$
= 299,792,458 m / 85,510.24 m $\rightarrow$

$D_{c\_toi178e}$ = 3,505 (exactly 3,505.92)

Therefore TOI-178 e planet's variable radius $R_{toi178e}$ becomes:

$R_{toi178e}$ = [GM_{toi178}/c^2] x $D_{c\_toi178e}^2$
= 959.80663634384929 m x $3,505^2$ →

$R_{toi178e}$ = 11,791,248,522 m or 11,791,248 km

which tightly matches TOI-178 e planet's observed orbital radius value as: 11,713,680 km (Δ=0.662%)

Orbital Velocity-Radius Verification Case Of Planet TOI-178 f:

With TOI-178 f planet's observed orbital velocity value as:

$V_{toi178f}$ = 74.20937 km/s or 74,209.37 m

based on values of:

TOI-178 f planet's semi-axis = 15,543,440 km (or 0.1039 AU)
(→ circumference = 97,662,313.83 km)

TOI-178 f planet's sidereal orbital period = 15.231915 days (or 1,316,037.45 seconds)

TOI-178 f planet's orbital quantizer $D_{ctoi178f}$ value becomes:

$D_{c\_toi178f}$ = c / $V_{toi178f}$
= 299,792,458 m / 74,209.37 m →

$D_{c\_toi178f}$ = 4,039 (exactly 4,039.81)

Therefore TOI-178 f planet's variable radius $R_{toi178f}$ becomes:

$R_{toi178f}$ = [GM_{toi178}/c^2] x $D_{c\_toi178f}^2$
= 959.80663634384929 m x $4,039^2$ →

$R_{toi178f}$ = 15,657,825,717 m or 15,657,825 km

which tightly matches TOI-178 f planet's observed orbital radius value as: 15,543,440 km (Δ=0.735%)

Orbital Velocity-Radius Verification Case Of Planet TOI-178 g:

With TOI-178 g planet's observed orbital velocity value as:

$V_{toi178g}$ = 66.97894 km/s or 66,978.94 m

based on values of:

TOI-178 g planet's semi-axis = 19,074,000 km
(→ circumference = 119,845,476.54 km)

TOI-178 g planet's sidereal orbital period = 20.7095 days

(or 1,789,300.8 seconds)

TOI-178 g planet's orbital quantizer $D_{c\_toi178g}$ value becomes:

$D_{c\_toi178g}$ $= c / V_{toi178g}$
                    $= 299{,}792{,}458 \text{ m} / 66{,}978.94 \text{ m}$ →

$D_{c\_toi178g}$ $= 4{,}475$ (exactly 4,475.92)

Therefore TOI-178 g planet's variable radius $R_{toi178g}$ becomes:

$R_{toi178g}$ $= [GM_{toi178}/c^2] \times D_{c\_toi178g}^2$
                 $= 959.80663634384929 \text{ m} \times 4{,}475^2$ →

$R_{toi178g}$ $= 19{,}220{,}727{,}771 \text{ m}$ or $19{,}220{,}727 \text{ km}$

which tightly matches TOI-178 g planet's observed orbital radius value as: 19,074,000 km (Δ=0.769%)

Summary Of Orbital Velocity-Radius-Period-Precession Quantizer Mechanism

Reason-For-Being Of Orbital Velocity-Radius-Period-Precession Quantizer Mechanism

All my findings so far and their demonstrations appear to confirm

> The existence of a universal quantizer mechanism that is capable of generating the orbital radius, orbital velocity, orbital period and orbital precession for all captives inside a cosmic system in a unified and harmonious manner. These cosmic systems can be galactic systems, star systems, planetary systems and black hole systems.

The said universal quantizer mechanism can be defined so far as the

> Orbital velocity-radius-period- precession quantizer mechanism

But, based on my analysis,

> The said orbital quantizer mechanism might host additional quantum characteristics that are still waiting to be discovered.

Orbital Quantizer's Reason-For-Being

Based on all my findings so far,

> The orbital velocity-radius-period-precession quantizer mechanism relies on a fundamental quantum component called:
>
> **Orbital Quantizer**

The orbital quantizer hereby is defined by

> A single and simple equation as:

$$Dc = [c / V] \qquad (=s1)$$

> where
> Dc is the orbital quantizer of a given captive (like a planet with respect to a star).

c is the speed of light
V is the actual orbital velocity of this captive

The said orbital quantizer equation is fundamental for the entire mechanism because:

Once a value of orbital quantizer is chosen, it becomes the core component that will be used to calculate and determine the orbital velocity, orbital radius, orbital period and orbital precession of the chosen captive.

As all my findings so far have revealed that,

For any captive, its orbital quantizer (Dc) determines its orbital radius (R) via the equation:

$$R = [GM/c^2] \times Dc^2 \qquad (=y2)$$

where:

Dc is the orbital quantizer for a specific captive
M is the mass of the captor
G is Newton's gravitational constant

GM/c^2 is the quantum core radius for the captive, which is also the value of Eg_1: the remaining orbital motion energy potential from the captor of the involved cosmic system (star systems and below)

For any captive, its orbital quantizer (Dc) also determines its orbital period (T) via the equation:

$$T = 2\pi \times ([GM/c^3] \times Dc^3) \qquad (=j2)$$

For any captive, its orbital orbital quantizer (Dc) also determines its orbital perihelion precession angle (ε) via the equation:

$$\varepsilon = (6\pi / Dc^2) / (1-e^2) \qquad (=d1.0)$$

All the equations of this orbital quantizer mechanism, as presented above, can demonstrate that

A captor is allowed to have millions of captives orbiting harmoniously around it – without possible collisions – as long as each of its captive is assigned with a unique value of orbital quantizer, which can be assimilated to a specific frequency.

Orbital Quantizer's Apparent Link With Lorentz Factor

Based on my previous findings,

> The discovered orbital quantizer appears to show up not only in the determination of the orbital velocity, orbital radius, orbital period and orbital precession of a captive, but also in the Lorentz factor's equation.

And the link between the orbital quantizer and the Lorentz factor has been demonstrated as follows:

The β ratio component of the Lorentz factor:

$$\beta = V / c$$

is same as:

$$\beta = 1 / D_c \qquad (=s6)$$

because:

$$D_c = c / V \qquad (=s1)$$

Therefore, the Lorentz factor's equation:

$$\gamma = 1/\sqrt{(1 - v^2/c^2)}$$

Hence its orbital quantizer-based alternative equation:

$$\gamma = 1/\sqrt{(1 - 1/D_c^2)} \qquad (s7)$$

with:

$$D_c = 1 / \beta \qquad (s6.1)$$

The alternative equation above (s7) shows the presence of the orbital quantizer inside the Lorentz factor. Such a presence corroborates the role, hence the reason-for-being, of the discovered orbital velocity-radius-period-precession quantizer mechanism in a larger scope. The additional scope occurs whenever an object moves inside a gravitational field of another object.

Roles Of Orbital Quantizer For Orbital Radius, Orbital Velocity, Orbital Period, Orbital Precession And Lorentz Factor

Based on all my present and past findings,

    The roles of the orbital quantizer can be summarized as:

        Determination of final orbital radius of a captive

        Determination of final orbital velocity of a captive

        Determination of final orbital period of a captive

        Determination of final orbital precession of a captive

        Determination of Einstein's special relativity effects known as time dilation and length contraction.

The said different roles of the orbital quantizer hereby reveals the quantum nature of the discovered orbital velocity-radius-period-precession quantizer mechanism.

Reasoning Path Leading To All Equations Of Orbital Velocity-Radius-Period-Precession Quantizer Mechanism

This chapter will explain

> The way how my additional discovered orbital quantizer mechanism related to orbital periods of captives comes about.

The outcome of this additional discovery is

> A set of equations that can calculate and predict the orbital period for any captive, described as follows:

$$T = 2\pi \times ([GM/c^3] \times Dc^3) \qquad (=j2)$$

> where:
> M is the mass of the captor.
> Dc is the orbital quantizer created by the said captor for a specific captive.
> G is Newton's gravitational constant
> c is the speed of light in vacuum

$$T / Dc^3 = [GM/c^3] \times 2\pi \qquad (=j3)$$

$$T \times c / 2\pi = [GM/c^2] \times Dc^3 \qquad (=j1)$$

These equations started out with my analysis of Kepler's third law from a new angle.

Kepler's Third Law Postulate

Kepler's Third Law Definition:

The crux of Kepler's third law says

> The square of the orbital period of any planet is proportional to the cube of the semi-major axis of its orbit.

Newton's Version Of Kepler's Third Law:

Newton had created his own version of Kepler's third law via the following equations (t0) and (j2.7):

$$T^2 = [4\pi^2 / G(M_1+M_2)] \times (R^3) \qquad (=t0)$$

$$T^2 = [4\pi^2 / GM] \times (R^3) \qquad (=j2.7)$$

where:
T is the orbital period of the captive.

a is the semi-major axis of the elliptical orbit of the captive.
G is Newton's gravitational constant
M1 is the mass of the captor
M2 is the mass of the captive

In the case where the captor's mass is highly dominant (like the Sun in the Solar system) then M2 can be ignored.

From Newton's Version Of Kepler's Third Law Equation To Speed-Of-Light-Based Orbital Period Quantizer Equation

My discovery of the said equation of orbital period quantizer mechanism starts with:

Rewrite Newton's version equation of Kepler's third law that has the orbital period as the central term:

$$T^2 = [4\pi^2 / GM] \times (a^3) \qquad (t0)$$

Into another equation that has this time the semi-major axis "a" as the central term:

$$a^3 = GM \times (T^2 / 4\pi^2) \qquad (t1)$$

Based on my previous discovery of the quantum equation of classical orbital velocity:

The orbital velocity of a captive is equationted as:

$$V = c / Dc \qquad (=s5)$$

Where:
c is the speed-of-light in vacuum
Dc as orbital quantizer (Dc = c/ V)

And based on the basic principles of geometry:

When an ellipse tends to become a circle. the semi-major axis "a" tends to become a radius, therefore:

a = orbital radius R = circumference / 2π

In other words, a semi-major axis "a" as a radius is equivalent to a [circumference / 2π].

Now let's go through step by step the transformation of the Newton's version equation (t1) of Kepler's third law via a set of consecutively rewritten transitional equations, presented hereafter:

Because the orbital period T is the duration that a test particle takes while moving along the orbit of a captive at a velocity V to achieve one full orbital revolution, then we must have:

$$a = [T \times V] / 2\pi$$

As a result, the rewritten equation of Kepler's third law

$$a^3 = GM \times (T^2 / 4\pi^2) \qquad (=t1)$$

must be equivalent to

$$([T \times V] / 2\pi)^3 = GM \times (T/2\pi)^2 \qquad (t2)$$

Because both left-hand and right-hand sides of the equation (t2) contains a common term "T/2π" then we can separate the latter from all other terms via the new transitional equation:

$$([T/2\pi] \times V)^3 = GM \times (T/2\pi)^2 \qquad (t3)$$

As the classical orbital velocity

$$V = c / Dc \qquad (=s5)$$

Then the transitional equation (t3) becomes:

$$([T/2\pi] \times [c/Dc])^3 = GM \times (T/2\pi)^2 \qquad (t4)$$

By combining "c" and "T" into one term "Txc", the transitional equation (t4) at this step becomes:

$$([T \times c] / [2\pi \times Dc])^3 = GM \times (T/2\pi)^2 \qquad (t5)$$

By applying first the quotient-to-power rule then second the product-to-power rule to all exponents inside the transitional equation (t5), we obtain a new transitional equation:

$$[T^3 \times c^3] / [(2\pi)^3 \times Dc^3] = [GM \times T^2] / (2\pi)^2 \qquad (t6)$$

Because "$(2\pi)^3$" is same as "$(2\pi)^{2+1}$" and "T^3" is same as "T^{2+1}", the transitional equation (t6) can be rewritten as:

$$[T^2 \times T \times c^3] / [(2\pi)^2 \times (2\pi) \times Dc^3] = [GM \times T^2] / (2\pi)^2 \qquad (t7)$$

Because the left-hand and right-hand sides of the transitional equation (t7) share the same terms "T^2" and "$(2\pi)^2$", and as these terms can cancel out each other, the transitional equation (t7) can be shortened in a new transitional equation as:

$$[T \times c^3] / [(2\pi) \times Dc^3] = GM \qquad (t8)$$

As $c^3 = $ "c^{2+1}", we can extract the sub-term "c^2" then move this sub-term from the left-hand side to the right-hand side of the transitional equation (t8) then obtain:

$$\mathbf{T \times c / [(2\pi) \times Dc^3] = [GM/c^2]} \qquad (t9)$$

The transitional equation (t9) reveals the gravitational energy quanta $[GM/c^2]$. This is the first hint of the presence of a gravitational energy as an actor in the determination of the orbital period of a captive via the orbital quantizer.

By moving the orbital quantizer "Dc^3" term from the left-hand side to the right-hand side of the transitional equation (t9), we get a new transitional equation:

$$T \times c / 2\pi = [GM/c^2] \times Dc^3 \qquad (=j1)$$

And the last transitional equation is no other than the discovered equation (j1) of the orbital period quantizer mechanism.

We can see here how this discovered equation of the orbital period quantizer mechanism came about.

From Speed-Of-Light-Based Orbital Period Quantizer Equation To Orbital Radius Quantizer Equation

Based on all my findings so far,

> The orbital quantizer mechanism does not only affect the quantum determination of the orbital period of every captive but also the latter's orbital radius.

Here is how the said mechanism link comes about,

Let's reequationte the equation (j1)

$$T \times c / 2\pi = [GM/c^2] \times Dc^3 \qquad (=j1)$$

By transferring one value "Dc" from the right-handed side of the equation to the left-handed side thereof, then we get a new transitional equation:

$$[T \times c / 2\pi] / Dc = [GM/c^2] \times Dc^2$$

which turns out to be equivalent to:

$$T \times [c / Dc] \times [1/2\pi] = [GM/c^2] \times Dc^2$$

Because the term "c/Dc" is the equation of the orbital velocity "V", hence:

$$[T \times V] / 2\pi = [GM/c^2] \times Dc^2$$

Because "T x V" is the circumference of the orbit of the captive, therefore "[TxV]/2π" is no other than the orbital radius R of this captive. The transitional equation at this step gets transformed one more time into:

$$R = [GM/c^2] \times Dc^2 \qquad (=y2)$$

This final equation suddenly looks exactly like the equation of the orbital quantizer mechanism for orbital radii of captives:

$$R = [GM/c^2] \times Dc^2 \qquad (=y2)$$

We can see here that,

The discovered orbital quantizer mechanism yields one single orbital quantizer value to determine three orbital features of a same captive: orbital period, orbital radius and orbital velocity.

The reasoning path hereby shows that the discovered orbital quantizer mechanism for the orbital period of a given captive leads to the latter's orbital radius. Furthermore, this reason path also shows that the said orbital quantizer mechanism yields one single orbital quantizer value to determine three orbital features of a same captive: orbital period, orbital radius and orbital velocity.

From Orbital Radius Quantizer Equation To Speed-Of-Light-Based Orbital Period Quantizer Equation

In this segment, we will see the reasoning path of

How the orbital quantizer dedicated to the orbital radius of a captive leads to the same orbital quantizer dedicated to its orbital period.

The reasoning path hereby starts with

Reequationting the equation (y2), presented hereafter:

$$R = [GM/c^2] \times Dc^2 \qquad (=y2)$$

By multiplying both sides of the equation by one "Dc". The outcome therefrom will be the transitional equation:

$$R \times Dc = [GM/c^2] \times Dc^3$$

As Dc is same as "c/v", we can rewrite the current transitional equation as:

$$R \times c/v = [GM/c^2] \times Dc^3$$

By switching the location of "v" upside down (on the right-handed side of the equation) we get:

$$(R/v) \times c = [GM/c^2] \times Dc^3$$

As the duration term "R/v" is same as "T/2π", we get:

$$(R/v) \times c = (T/2\pi) \times c = [GM/c^2] \times Dc^3 \quad \rightarrow$$

$$T \times c / 2\pi = [GM/c^2] \times Dc^3 \qquad (=j1)$$

We can see here that,

> The reasoning path hereby shows that the discovered orbital quantizer mechanism for the orbital radius of a given captive leads to the latter's orbital period. Furthermore, this reason path also shows that the said orbital quantizer mechanism yields one single orbital quantizer value to determine three orbital features of a same captive: orbital period, orbital radius and orbital velocity.

Discovery Of Orbital Period Creation's Quantizer Mechanism

In this chapter, we will see how

> A quantizer mechanism can arise from Newton-Kepler law of orbital period of celestial objects when we introduce the orbital quantizer term into the said law's equation.

First from the transitional equation (t9):

$$T \times c / [(2\pi) \times Dc^3] = [GM/c^2] \qquad (=t9)$$

And by moving the term "c" from the left-handed side to the right-handed side of the transitional equation at this step, we obtain c^3 from c^2 as follows:

$$T / [(2\pi) \times Dc^3] = [GM/c^3] \qquad (t10)$$

Inside the transitional equation (t10), appears a new gravitational energy that determines the orbital period as:

$$[GM/c^3] \qquad (t11)$$

The transitional equation (t10) can be rewritten as:

$$T = 2\pi \times ([GM/c^3] \times Dc^3) \qquad (=j2)$$

which reveals that the orbital period of a captive is made of a special gravitational energy quanta "GM/c^3" and a specific orbital quantizer Dc to the power of 3.

or:

$$T / Dc^3 = [GM/c^3] \times 2\pi \qquad (=j3)$$

Discovery Of Orbital Velocity-Radius-Period-Precession Quantizer Mechanism Generating Discrete Not Random Orbital Period Jumps

Based on my findings,

The equation of orbital quantizer mechanism for orbital periods of captives:

$$T / Dc^3 = [GM/c^3] \times 2\pi \qquad (=j3)$$

reveals a hidden quantum mechanism but very relevant to the potential presence of a jump between different gravitational energy levels inside this equation.

And here is why:

Based on the equation (j3),

The right-handed side of the equation (j3) always yields the same constant value that depends only on the gravitational mass "GM" of the captor.

As a result, only the terms "Dc" and "T" inside the left-handed side of the equation (j3) can change their value; however their changes of values are interdependent because they must always yield the same said constant value "$[GM/c^3] \times 2\pi$" of the right-handed side of the equation (j3).

In other words, if the Dc value is increased (which leads to decrease of orbital velocity), the T value must be proportionally increased too, and vice versa.

Same Gravitational Energy Quanta From Orbital Velocity-Radius-Period-Precession Quantizer Mechanism Can Generate Different Discrete Orbital Period Jumps:

Let's examine the equation (j3) again:

By reversing upside down both the left-handed side and the right-handed side of the equation (j3), we get:

$$T / Dc^3 = [GM/c^3] \times 2\pi \qquad (=j3)$$

$\rightarrow$

$$Dc^3 / T = 1 / ([GM/c^3] \times 2\pi) \qquad (j3.1)$$

By multiplying and dividing only the left-handed side of the equation (j3.1) by an identical orbital quantizer change ratio (Δdc) raised to the power of 3, we get a new equation:

$$\mathbf{Dc^3 \times \Delta dc^3 / T \times \Delta dc^3 = 1 / ([GM/c^3] \times 2\pi)} \qquad (j3.2)$$

which is equivalent to

$$\mathbf{(Dc \times \Delta dc)^3 / T \times \Delta dc^3 = 1 / ([GM/c^3] \times 2\pi)} \qquad (j3.3)$$

It becomes clear that the value of the orbital quantizer change ratio (Δdc) can take any value greater than zero.

The equation (j3.2) shows that the captor must generate orbital periods in discrete values for successive captives orbits, based on the cube value of a given orbital quantizer (Dc $\times$ Δdc).

The last equation (j3.2) also shows that the captor uses the same gravitational energy quanta ($\mathbf{GM/c^3}$) to generate discrete orbital periods, and does not need any extra mass to generate orbits of faraway captives.

Orbital Velocity-Radius-Period-Precession Quantizer Mechanism Reveals Quantum Root Of Newton's law of orbital velocity "V2 = GM/R":

Based on my findings,

The equation (j3) also reveals the deep quantum root of Newton's law of orbital velocity "$V^2 = GM/R$".

And here is why,

According to this Newton's law "$V^2 = GM/R$",

when the orbital velocity V decreases, the orbital radius R must increase proportionally.

According to the discovered orbital velocity-radius-period-precession quantizer mechanism,

$$Dc = c / V \qquad (=s1)$$

Therefore the decrease of the orbital velocity V from this Newton law leads to the increase of the Dc value.

Therefore inside the equation (j3)

$$T / Dc^3 = [GM/c^3] \times 2\pi \qquad (=j3)$$

the decrease of the orbital velocity V from this Newton law must ultimately lead to the increase of the Dc value to the power of 3.

And inside the equation (j3) the said increase of the Dc value must translate itself into a proportional increase of the orbital period T of value, hence the proportional increase of the orbital radius of the captive, based on the equation:

$$R = [GM/c^2] \times Dc^2 \qquad (=y2)$$

And this equation (y2) is from the discovered orbital velocity-radius-period-precession quantizer mechanism,

Let's examine the case of

A hypothetical slowest orbital velocity of 1m/second for a hypothetical planet around the Sun.

Dc_slowplanet = 299,792,458 m / 1 m = 299,792,458

$GMsun/c^3$ = 4.92571420697396808e-6 m

based on the values of:

Msun = 1.9885e+30 kg (as mass of the Sun)
G = 6.6743e-11 (N kg^{-2} m²)
c = 299,792,458 m

From the equation

$$Tslowplanet / Dc\_slowplanet^3 = [GMsun/c^3] \times 2\pi \qquad (=j3)$$

Let's calculate

$T_{slowplanet}$ = $[GM_{sun}/c^3]$ x 2π x $Dc_{slowplanet}^3$
= 4.92571420697396808e-6 secs x
2π x 2.69440024173739895593359 12e+25

Hence:

$T_{slowplanet}$ = 833,894,649,589,167,778,346 seconds
or 26,442,625,874,846.773438 Earth years

Because the hypothetical orbital velocity is 1m/second therefore the hypothetical orbital circumference must be
833,894,649,589,167,778,346 m

That leads to a hypothetical orbital radius of
132,718,455,499,999,999,999 m
(= circumference/2π)
or 1.32718455499999999999e+20 m

Because the hypothetical orbital velocity is 1m/second therefore the hypothetical orbital circumference ($Circ_{slowplanet}$) must be:

$Circ_{slowplanet}$ = 833,894,649,589,167,778,346 m

That leads to an hypothetical orbital radius of $R_{slowplanet}$ must be:

$R_{slowplanet}$ = 132,718,455,499,999,999,999 m
(= $Circ_{slowplanet}/2\pi$)
or ~1.327184555e+20 m

From Newton's orbital velocity law equation:

$R_{slowplanet}$ = GM_{sun} / $V_{slowplanet}^2$
= [6.6743e-11 x 1.9885e+30] / 1^2
= 1.327184555e+20 m
or 1.327184555e+17 km

We can conclude here that the equation (j3) yields the same value as Newton's orbital velocity law equation.

Let's examine the case of

A hypothetical slowest orbital velocity of 2m/second for a hypothetical planet around the Sun.

Dc_slowplanet = 299,792,458 m / 2 m = 149,896,229

$GM_{sun}/c^3 = 4.92571420697396808\text{e-}6$ m

    based on the values of:

    $M_{sun} = 1.9885\text{e+}30$ kg (as mass of the Sun)
    $G = 6.6743\text{e-}11$ (N kg^{-2} m^2)
    $c = 299,792,458$ m

From the equation

    $T_{slowplanet} / D_{c\_slowplanet}^3 = [GM_{sun}/c^3] \times 2\pi$ (=j3)

Let's calculate

    $T_{slowplanet}$ = $[GM_{sun}/c^3] \times 2\pi \times D_{c\_slowplanet}^3$
                   = $4.92571420697396808\text{e-}6$ secs x
                   $2\pi \times 3.36800030217174 8692416989\text{e+}24$

Hence:

    $T_{slowplanet}$ = 104,236,831,198,645,972,293 seconds
                   or 3,305,328,234,355 Earth years

Because the hypothetical orbital velocity is 2m/second therefore the hypothetical orbital circumference Circ$_{slowplanet}$ must be:

    Circ$_{slowplanet}$ = 208,473,662,397,291,944,586 m

That leads to an hypothetical orbital radius of R$_{slowplanet}$ must be:

    R$_{slowplanet}$ = 33,179,613,875,000,000,000 m
               (= Circ$_{slowplanet}/2\pi$)
               or $3.3179613875\text{e+}19$ m

From Newton's orbital velocity law equation:

    R$_{slowplanet}$ = $GM_{sun} / V_{slowplanet}^2$
               = $[6.6743\text{e-}11 \times 1.9885\text{e+}30] / 2^2$
               = $3.3179613875\text{e+}19$ m
               or $3.3179613875\text{e+}16$ km

We can conclude here that the equation (j3) yields the same value as Newton's orbital velocity law equation.

Basic Star System's Orbital Quantizer Constraints

Based on my analysis,

All the equations that yield the orbital quantizer that determines the set of orbital velocity, orbital radius and orbital period for a given captive must respect the following constraints:

Orbital quantizer of the closest planet of a given star system cannot yield an orbital distance smaller than the radius of the star's core.

Orbital quantizer of any planet of a given star system cannot yield an orbital distance smaller than the orbital distance of the preceding planet (from the star outwards) + the orbital distance of the latter's farthest captive moon or asteroids.

Here are a sample of orbital quantizer constraints in the Solar system:

Orbital quantizer constraint imposed by the Sun's Core:

Orbital Quantizer Occupied by Sun's Core	471,121
Based on:	
Sun's core radius	695,700 km
Sun's GM/c^2 (GM_{sun}/c^2)	1,476.6919 m

The actual orbital quantizer of Mercury demonstrates that the quantum engine of the Sun cannot directly use the value of the found orbital quantizer for a specific planet but must instead use the square value thereof. Here is why:

Mercury's Dc (Discovered value of Dc_mercury)	6,262.62
Is smaller than	
Sun core's Dc	471,121
But	
Mercury's Dc^2 ($Dc\_mercury^2$)	39,220,409
Is much bigger than Sun core's Dc	

Discovery Claim Of Orbital Quantizer Behind Captives' Perihelion Precession Mechanism

Simpler Alternative To Gerber-Einstein Equation Of Perihelion Precession Angle

Based on my previous findings,

The discovered orbital quantizer mechanism appears to show up also in the determination of the perihelion precession of captives.

My two discovered equivalent orbital quantizer-based equations for the angle (ε) of perihelion precession of any captive:

$\varepsilon = (6\pi / Dc^2) / (1-e^2)$ (d1.0)

and

$\varepsilon = (6\pi \times [v/c]^2) / (1-e^2)$ (d0.0)

where:
ε is the angle of perihelion precession of a given captive
Dc is the orbital quantizer of the said captive.
v is the orbital velocity of the said captive
c is the speed of light in vacuum

yield exactly the same result as Einstein's equation for the angle (ε) of perihelion precession of any captive:

$$\varepsilon = 24\pi^3 \frac{a^2}{T^2 c^2 (1-e^2)}$$

There is an interesting fact that

Paul Gerber had come up with the exact equation roughly two decades before, with a reasoning and an interpretation that are different from Einstein's ones: spacetime paradigm and general relativity theory.

And because of Paul Gerber's antecedence, it would be more fair to call this equation "Gerber-Einstein Equation".

Discovery Of Hidden Orbital Quantizer Inside Gerber-Einstein Equation Of Captives' Perihelion Precession Angle

Let's go through and examine

The Gerber-Einstein equation that calculates the angle (ε) of perihelion precession of any captive:

$$\varepsilon = 24\pi^3 \times [a^2/T^2c^2(1-e^2)] \tag{e1}$$

where:
T is the orbital period of the captive's orbital revolution.
a is the semi-major axis (aka orbital radius) of the captive.
c is the speed of light in vacuum
e is the eccentricity of the (elliptical) orbit of the captive

For simplification of the study,

Let's assume that the orbit of the captive is a circle, therefore any reference to an ellipse can be ignored. Because the equation component "$(1-e^2)$" is linked to orbit eccentricity therefore can be removed temporarily from the equation. As a result, the equation can be rewritten as:

$$\varepsilon = 24\pi^3 \times [a^2/T^2c^2] \tag{e2}$$

The equation (e2) can be rewritten as:

$$\varepsilon = 6\pi \times 4\pi^2 \times [a^2/T^2c^2] \tag{e3}$$

And the regrouped term "$4\pi^2 \times [a^2/T^2c^2]$" is no other then $[2\pi a/Tc]$ raised to the power of 2, hence:

$$[2\pi a/Tc]^2 = 4\pi^2 \times [a^2/T^2c^2] \tag{e4}$$

Then the equation (e3) becomes:

$$\varepsilon = 6\pi \times [2\pi a/Tc]^2 \tag{e5}$$

Let's imagine a hypothesis in which

Every captive has a virtual twin that revolves along the same orbit around their shared captor, but exceptionally the virtual twin moves at the speed of light.

In this hypothesis,

For the virtual twin, we have:

Tc = Orbital period of the real captive in seconds (T) x speed of light (c)

hence:

Tc = Total distance made by the virtual twin when the real captive finishes one orbit around the captor.

For the real captive, we have instead:

TV = Orbital period of actual captive in seconds (T) x its actual orbital velocity (V)

hence:

TV = Total distance made by the real captive after finishing one orbit around the captor.

It turns out that

The term "TV" must be no other than the length of the real captive's orbit (presented by "$2\pi a$").

Therefore, the transitional equation (e5):

$$\varepsilon = 6\pi \times [2\pi a / Tc]^2 \qquad (=e5)$$

becomes:

$$\varepsilon = 6\pi \times [TV/Tc]^2 \qquad (e6)$$

And it turns out also that

The term "Tc" must be equal to the "number of repeated orbits" made by the virtual twin, and this number of repeated orbits must be equal to the ratio between the speed of light and the actual velocity of the real captive.

In this transitional equation (e6), the orbital period T is both present at the top and the bottom of the term "TV/Tc", therefore can be removed from the equation, hence:

$$\varepsilon = 6\pi \times [V/c]^2 \qquad (e7)$$

Thanks to the core orbital quantizer equation (Dc) from my discovered orbital velocity-radius-period-precession quantizer mechanism:

$$Dc = c / V \qquad (=s1)$$

where:

V is the actual orbital velocity of the captive (present in Newton's equation of orbital velocity)
c is the speed-of-light in vacuum

the equation (e7) can be rewritten as:

$$\varepsilon = 6\pi \times [V/c]^2 = 6\pi \times [(c/Dc)/c]^2 = 6\pi \times (1/Dc)^2$$

Hence:

$$\boldsymbol{\varepsilon = 6\pi \times 1/Dc^2 = 6\pi/Dc^2} \tag{=e8}$$

Here it becomes clear that

> Gerber-Einstein's equation of captives' perihelion precession angle has a simple alternative incarnated by the equation (e8).

And this alternative reveals that

> The discovered orbital velocity-radius-period-precession quantizer mechanism is also present in the perihelion precession mechanism in which all captives operate.

Working Scientific Data Set

All calculations made in this book are based on the set of scientific data collected from various difference public sources, as described in this chapter.

Some scientific feature values may have different values as their observations could not be firmly established as of 2022. In these cases, the most widely accepted values will be used.

Basic Constants

Speed of Light in Vacuum:

c = 299,792,458 m/s

Parsec Length:

Parsec = 3.09e+13 km (or 3.09×10^{13} km)
  or exactly 30,856,775,814,672 km

Kiloparsec or Kpc = 3.09e+16 km (or 3.09×10^{16} km)

Megaparsec or Mpc = 3.09e+19 km (or 3.09×10^{19} km)
(= million parsecs)

G Constant (Newton's universal gravitational constant):

G = 6.6743e-11 N kg^{-2} m² (or 6.6743×10^{-11})

AU (Astronomical units):

AU = 149.6e+6 km (or 149.6×10^{6} km)
  or 149,597,870.7000 km (exact value)

Sun's Mass values:

1.9885e+30 kg (or 1.9885×10^{30} kg)

Sun's GM (geocentric gravitational constant μ) value:

GMsun = 1.327184555e+20 m³ s⁻²
(= 6.6743e-11 kg⁻²m² x 1.9885e+30 kg)

Earth's Mass value:

Mearth = 5.9724e+24 kg (or 5.9724 x 10^{24} kg)

Studied Solar Planets' Orbital Period Values

(From Wiki source)

Mercury's Sidereal Orbital Period	87.9691 days (or 7,600,530.24 seconds)
Venus' Sidereal Orbital Period	224.7 days (or 19,414,080 seconds)
Earth's Sidereal Orbital Period	365.25636 days (or 31,558,149.504 seconds)
Mars' Sidereal Orbital Period	687.0 days (or 59,360,000 seconds)
Jupiter's Sidereal Orbital Period	4,331 days (or 374,198,000 seconds)
Saturn's Sidereal Orbital Period	10,747 days (or 928,540,800 seconds)
Uranus' Sidereal Orbital Period	30,589 days (or 2,642,890,000 seconds)
Neptune's Sidereal Orbital Period	59,800 days (or 5,166,720,000 seconds)

Studied Solar Planets' Mean Orbital Velocity Values

(From Wiki source)

Mercury's Mean Orbital Velocity	47.87 km/sec (or 47,870 m/sec)
Venus' Mean Orbital Velocity	35.02 km/sec (or 35,020 m/sec)

Earth's Mean Orbital Velocity	29.78 km/sec (or 29,780 m/sec)
Mars' Mean Orbital Velocity	24.077 km/sec (or 24,077 m/sec)
Jupiter's Mean Orbital Velocity	13.07 km/sec (or 13,070 m/sec)
Saturn's Mean Orbital Velocity	9.69 km/sec (or 9,690 m/sec)
Uranus' Mean Orbital Velocity	6.81 km/sec (or 6,810 m/sec)
Neptune's Mean Orbital Velocity	5.43 km/sec (or 5,430 m/sec)

Studied Solar Dwarf Planets' Mean Orbital Velocity Values

(From Wiki source)

Mathilde's Mean Orbital Velocity	17.98 km/sec (or 17,980 m/sec)
Juno's Mean Orbital Velocity	17.93 km/s
Eugenia's Mean Orbital Velocity	18.06 km/s
Ceres' Mean Orbital Velocity	17.87 km/s
Pallas' Mean Orbital Velocity	

Studied Planets' Orbital Radius Values

Planets' Orbital radius is also called mean distance from planet to host star.

In the calculation of Kepler's laws, the semi-major axis of elliptical orbit is used because all planetary orbits are elliptical.

By definition:
- A Semi-major axis is the longest radius of an ellipse.
- A Semi-minor axis is the shortest radius of an ellipse.

Because some planets may have contradicting values, each radius value is provided with data sources.

Mercury's Orbital Radius (From NASA data + others)	57,910,000 km
Venus' Orbital Radius (From Wiki data)	108,200,000 km
(From NASA data)	108,209,475 km
Earth's Orbital Radius (From NASA data)	149,598,000 km
Mars' Orbital Radius (From NASA data as 1.52366231 AU) (confirmed by Wiki data)	227,936,637 km
Mathilde's Orbital Radius: (From Wiki data as 2.648402147 AU) (considered as dwarf planet)	396,200,961 km
Juno's Orbital Radius: (From NASA data as 2.67070 AU) (considered as dwarf planet)	399,536,720 km
Eugenia's Orbital Radius: (From Wiki data as 2.7199 AU)	406,897,040 km
Ceres' Orbital Radius: (From NASA data as 2.768 AU)	414,092,800 km
Pallas' Orbital Radius: (From NASA data as 2.772 AU)	414,691,200 km
Jupiter's Orbital Radius (From NASA data as 5.20336301 AU) (Other value of 778,479,000 km does not match its AU value)	778,412,027 km
Saturn's Orbital Radius (From NASA data as 9.53707032 AU) (NASA value of 1,432,041,000 km mismatched its AU value)	1,426,725,400 km
Uranus' Orbital Radius (From NASA data as 19.19126393 AU) (From WIKI data)	2,870,972,219 km
Neptune's Orbital Radius	4,498,252,910 km

(From NASA data as 30.06896348 AU)
Other value : 4,498,438,349 km

90482 Orcus   5,860,430,400 km
(From Wiki data as 39.174 AU)

Pluto's Orbital Radius   5,906,376,272 km
(From NASA data as 39.48168677 AU)
(confirmed by newworldencyclopedia.org)
Other value: 5,906,453,414 km
Other value: 5,906,522,160 km
(Princeton.edu as 39.4821 AU)
    5,906,438,090 km
(Biggest trans-Neptunian object)

Haumea's Orbital Radius   6,460,057,120 km
(From Princeton.eu data as 43.1822 AU)
(3rd Biggest trans-Neptunian object)

50000 Quaoar's Orbital Radius   6,537,520,000 km
(From ~43.7 AU)

Makemake's Orbital Radius   6,796,328,000 km
(From Princeton.eu data as 45.430 AU)
(4th Biggest trans-Neptunian object)

Eris' Orbital Radius   10,152,454,400 km
(From Princeton.eu data as 67.864 AU)
(2nd Biggest trans-Neptunian object)

2018VG18's Orbital Radius   12,221,272,800 km
(From Wiki data as 81.693 AU)
(aka ""Farout")

2005RM43's Orbital Radius   13,643,520,000 km
(From Wiki data as 91.2 AU)

2005SA278's Orbital Radius   13,897,840,000 km
(From Wiki data as 92.9 AU)

2000OM67's Orbital Radius   14,960,000,000 km
(From Wiki data as 100 AU)

2008 ST291's Orbital Radius   15,109,600,000 km
(From Wiki data as 101 AU)

2012 FL84's Orbital Radius   15,408,800,000 km
(From Wiki data as 103 AU)

1999 RZ215's Orbital Radius   15,558,400,000 km
(From Wiki data as 104 AU)

2005QU182's Orbital Radius (From Wiki data as 115 AU)	17,204,000,000 km
2014YK50's Orbital Radius (From Wiki data as 117 AU)	17,503,200,000 km
2009YD7's Orbital Radius (From Wiki data as 120 AU)	17,952,000,000 km
1999RD215's Orbital Radius (From Wiki data as 125 AU)	18,700,000,000 km
2000 PJ30's Orbital Radius (From Wiki data as 125 AU)	18,700,000,000 km
2014MJ70's Orbital Radius (From Wiki data as 126 AU)	18,849,600,000 km
2013GP136's Orbital Radius (From Wiki data as 151 AU)	22,589,600,000 km
2003 HB57's Orbital Radius (From Wiki data as 159 AU)	23,786,400,000 km
2015SO20's Orbital Radius (From Wiki data as 171 AU)	25,581,600,000 km
2005PU21's Orbital Radius (From Wiki data as 182 AU)	27,227,200,000 km
2007VJ305's Orbital Radius (From Wiki data as 201 AU)	30,069,600,000 km
2013UT15's Orbital Radius (From Wiki data as 204 AU)	30,518,400,000 km
2000CR195's Orbital Radius (From Wiki data as 212 AU)	31,715,200,000 km
2001FP185's Orbital Radius (From Wiki data as 212 AU)	31,715,200,000 km
2012VP113's Orbital Radius (From Wiki data as 271.5 AU) (From J. Becker et al 2019 as 318.0 AU) (Planet 9's revealer)	40,616,400,000 km
2010NV1's Orbital Radius (From Wiki data as 280 AU)	41,888,000,000 km
2014 LM28's Orbital Radius	43,384,000,000 km

(From Wiki data as 290 AU)

Void Orbital Radius Gap of 52 AU

Planet 9's Orbital Radius Range Start => 44,880,000,000 km
(300 AU)
(=380 AU -80/+140 AU)

2004VN112's Orbital Radius (From Wiki data as 332.80 AU) (aka "474640 Alicanto") (Planet 9's revealer)	49,786,880,000 km

Void Orbital Radius Gap of 31 AU

2013RF98's Orbital Radius (From Wiki data as 364.0 AU) (Planet 9's revealer)	54,454,400,000 km
2010GB174's Orbital Radius (From Wiki data as 371.0 AU) (Planet 9's revealer)	55,501,600,000 km
2009MS9's Orbital Radius (From Wiki data as 389.0 AU)	58,194,400,000 km

Void Orbital Radius Gap of 84 AU

2007TG422's Orbital Radius (From Wiki data as 473.0 AU) (From J. Becker et al 2019 as 482.0 AU in presence of Neptune) (Planet 9's revealer)	70,760,800,000 km

Void Orbital Radius Gap of 33 AU

Sedna's Orbital Radius (From ~506 AU)	75,697,600,000 km

Planet 9's Orbital Radius Range End => 77,792,000,000 km
(520 AU: from 380 AU -80/+140 AU)
(Estimated range value as of 2023)

2006 SQ372's Orbital Radius (From Wiki data as 1114.0 AU)	166,654,400,000 km
Leleākūhonua's Orbital Radius (From Wiki data as 1290 AU)	192,984,000,000 km
The Goblin's Orbital Radius (From ~ 1,369.79770882 AU)	204,918,820,941 km

Moon's Orbital Radius   384,399 km
  (From NASA data)
  (From WIKI data as 0.002569 AU)

Studied Planets' Core Radius

Sun's Core Radius (mean value)	695,700 km
Earth's Core Radius	6,378 km
Jupiter's Core Radius	69,911 km
Saturn's Core Radius	60268 km

Studied Planets' Orbital Circumference Values

The values, cited hereafter, are related to the orbital circumferences of the planets inside the Solar system, and come from most reliable scientific data sources. Only when such data does not exist then the usual "2πR" equation will be used to determine this value.

Mercury's Orbital Circumference: (vs. 2πR value of 363,859,261 km)	359,922,622 km
Venus' Orbital Circumference: (vs. 2πR value of 679,840,650 km)	679,892,378 km
Earth's Orbital Circumference: (from 2πR, as no official values exist)	939,951,955 km
Mars' Orbital Circumference: (from NASA data) or (from Wiki as 9.553 AU:1,429,108,458 km)	1,429,085,052 km
Ceres' Orbital Circumference: (vs. 2πR value of 414,092,800 km)	2,601,821,796 km
Jupiter's Orbital Circumference: (vs. 2πR value of 4,890,907,010 km)	4,887,595,931 km
Saturn's Orbital Circumference: (from planetary-science.org) (vs. 2πR value of 8,997,778,970 km)	8,957,504,604 km
Uranus' Orbital Circumference:	18,026,744,947 km

(vs. 2πR value of 18,036,429,382 km)

Neptune's Orbital Circumference: 28,263,782,131 km
(From planetary-science.org)
28,263,736,967 km
(vs. 2πR value of 28,264,521,739 km)

Pluto's Orbital Circumference: 36,530,016,653 km
(244.186 AU)
(vs. 2πR value of 36,880,136,337 km)

Sedna's Orbital Circumference: 475,618,278,197 km
(vs. 2πR value of 75,697,600,000 km

The Goblin's Orbital Circumference: 1,287,542,924,901 km
(vs. 2πR value of 204,918,820,941 km)

Moon's Orbital Circumference: 2,412,528 km
(from NASA data of velocity x orbital period =
1.022 km/sec x 27.3217 days)

Studied Planets' Orbital Eccentricity Values

Here is the list of eccentricity values "e" and its square value for each planet inside the Solar system:

Mercury Orbital Eccentricity 0.206
$e^2 = 0.042436$

Venus Orbital Eccentricity 0.00677
$e^2 = 0.000045$

Earth Orbital Eccentricity 0.0167
$e^2 = 0.00028$

Mars Orbital Eccentricity 0.0934
$e^2 = 0.0087$

Ceres Orbital Eccentricity 0.0758
$e^2 = 0.00574$

Jupiter Orbital Eccentricity 0.0487
$e^2 = 0.0023$

Saturn Orbital Eccentricity 0.0520
$e^2 = 0.0027$

Uranus Orbital Eccentricity	0.0469	
	$e^2 = 0.0022$	
Neptune Orbital Eccentricity	0.00858	
	$e^2 = 0.00007$	
Pluto Orbital Eccentricity	0.2444	
	$e^2 = 0.0597$	
Sedna Orbital Eccentricity	0.8496	
	$e^2 = 0.7218$	
The Goblin Orbital Eccentricity	0.94	
	$e^2 = 0.8836$	
Moon Orbital Eccentricity (of the Earth)		

Studied Sun and Planets' Mass Values

Sun's Mass	1.9885e+30 kg
Mercury's Mass	3.3022e+23 kg (3.3022 x 10^{23} kg)
→ Sun-Mercury mass ratio	6,021,743
Venus' Mass	4.8685e+24 kg (or 4.8685 x 10^{24} kg)
→ Sun-Venus mass ratio	408,442
Earth's Mass	5.9724e+24 kg (or 5.9724 x 10^{24} kg)
→ Sun-Earth mass ratio	332,948
→ Mercury mass ratio	18.08
Mars' Mass	6.4185e+23 kg (or 6.4185 x 10^{23} kg)
→ Sun-Mars mass ratio	3,098,075
Ceres' Mass	9.393e+20 kg
Jupiter's Mass	1.8986e+27 kg (or 1.8986 x 10^{27} kg)
→ Sun-Jupiter mass ratio	1,047
Saturn's Mass	5.6846e+26 kg (or 5.6846 x 10^{26} kg)

→ Sun-Saturn mass ratio	3,498
Uranus' Mass	8.6810e+25 kg
→ Sun-Uranus mass ratio	22,906
Neptune's Mass	10.243e+25 kg
→ Sun-Neptune mass ratio	19,413
Pluto's Mass	1.25e+22 kg (or 1.25×10^{22} kg)
→ Sun-Pluto mass ratio	1.27264×10^{-14}
Sedna's Mass	No reliable data available as of 2022
Moon's Mass (of the Earth)	7.346e+22 kg (or 7.346×10^{22} kg)

Studied Sun and Planet Core Radius Values

Sun core radius	
Mercury core radius	2,439.5 km
Venus core radius	6,052 km
Earth core radius	6,378 km
Mars core radius	3,396 km
Ceres core radius	
Jupiter core radius	71,492 km
Saturn core radius	60,268 km
Uranus core radius	25,559 km
Neptune core radius	24,764 km
Pluto core radius	1,188 km

Studied Solar Planets Orbital Resonances

(From source)

Solar Planets:	Resonances:

Neptune - Pluto   3:2

Studied Solar Asteroids' Observed Orbital Radius Values

Asteroid's Orbital Radius is defined as the mean distance from an asteroid to its host star; it is technically the semi-major axis of asteroid's elliptical orbit

The list of observed orbital radii of the studied asteroids of the Solar system is as followed:

Ra-shalom's Orbital Radius: (Full name Asteroid 2100 Ra-shalom) (From IAU data as 0.8320 AU)	124,467,200 km
Aten's Orbital Radius: (Full name Asteroid 2062 Aten) (From IAU Minor Planet center data as 0.9669 AU)	144,648,240 km
Icarus' Orbital Radius: (Full name Asteroid 1566 Icarus) (From WIKI data as 1.0781 AU)	161,283,760 km
Phaethon's Orbital Radius: (Full name Asteroid 3200 Phaethon) (From IAU Minor Planet center data as 1.2714 AU)	190,201,440 km
Toro's Orbital Radius: (Full name Asteroid 1685 Toro) (From IAU data as 1.3675 AU)	204,578,000 km
Apollo's Orbital Radius: (Full name Asteroid 1862 Apollo) (From IAU Minor Planet center data as 1.4703454 AU)	219,963,671 km
Adonis' Orbital Radius: (Full name Asteroid 2101 Adonis) (From IAU data as 1.8748413 AU)	280,476,258 km

Studied Solar Asteroids' Observed Orbital Period Values

The list of observed orbital periods of the studied asteroids of the Solar system is as followed:

Ra-shalom's Orbital Period:	0.759487 years (= 23,951,182 secs)
Aten's Orbital Period:	0.950 years (= 29,959,000 secs)
Icarus' Orbital Period:	1.119939 years (or or 408.778 days) (= 35,318,419.2 secs)
Phaethon's Orbital Period:	1.4346 years (= 45,241,545 secs)
Toro's Orbital Period:	1.599 years (= 50,464,598 secs)
Apollo's Orbital Period:	1.785 years (= 56,291,760 secs)
Adonis' Orbital Period:	2.567 years (= 80,952,912 secs)

Studied Solar Asteroids' Observed Orbital Circumference Values

All the following circumference values of asteroids are calculated from the "$2\pi R$" equation with their semi-major axis values:

Ra-shalom's Orbital Circumference: (Full name Asteroid 2100 Ra-shalom)	782,050,482 km
Aten's Orbital Circumference: (Full name Asteroid 2062 Aten)	908,851,696 km
Icarus' Orbital Circumference: (Full name Asteroid 1566 Icarus)	1,013,375,751 km
Phaethon's Orbital Circumference: (Full name Asteroid 3200 Phaethon)	1,195,070,893 km
Toro's Orbital Circumference: (Full name Asteroid 1685 Toro)	1,285,401,483 km

Apollo's Orbital Circumference: 1,382,072,505 km
(Full name Asteroid 1862 Apollo)

Adonis' Orbital Circumference: 1,762,284,303 km
(Full name Asteroid 2101 Adonis)

Studied Solar Asteroids' Observed Orbital Velocity Values

Asteroid orbital velocity values (V) here are deduced from the basic calculation method using observed orbital periods (T) and orbital circumference (Circ) or radius (R) ($V = T/Circ = T/Rx2\pi$).

The list of observed orbital velocities of the studied asteroids of the Solar system is as followed:

Ra-shalom's Orbital Velocity: 32.652 km/s
(=124,467,200 km x 2π / 23,951,182 secs)

Aten's Orbital Velocity: 30.336 km/s
(=144,648,240 km x 2π / 29,959,000 secs)

Icarus' Orbital Velocity: 28.69 km/s
(=161,283,760 km x 2π / 35,318,743 secs)

Phaethon's Orbital Velocity: 26.415 km/s
(=190,201,440 km x 2π / 45,241,545 secs)

Toro's Orbital Velocity: 25.471 km/s
(=204,578,000 km x 2π / 50,464,598 secs)

Apollo's Orbital Velocity: 24.552 km/s
(=219,963,671 km x 2π / 56,291,760 secs)

Adonis' Orbital Velocity: 21.769 km/s
(=280,476,258 km x 2π / 80,952,912 secs)

Studied Solar Asteroids' Observed Orbital Eccentricity Values

Ra-shalom's Orbital Eccentricity: 0.4365562
(Full name Asteroid 2100 Ra-shalom)

Aten's Orbital Eccentricity: 0.1828
(Full name Asteroid 2062 Aten)

Icarus' Orbital Eccentricity: 0.8270
(Full name Asteroid 1566 Icarus)

Phaethon's Orbital Eccentricity: 0.8897964
(Full name Asteroid 3200 Phaethon)

Toro's Orbital Eccentricity: 0.4359178
(Full name Asteroid 1685 Toro)

Apollo's Orbital Eccentricity: 0.5599167
(Full name Asteroid 1862 Apollo)

Adonis' Orbital Eccentricity: 0.7639970
(Full name Asteroid 2101 Adonis)

Studied Earth's Satellites' Values

Lageos II's Orbital Eccentricity: 0.004
(almost circular orbit)

Lageos II's Orbital Radius: 12,286 km
(semi-major axis)

This value is obtained based on 3 elements:
. Earth's radius of 6378 km
. Periapsis altitude of 5858 km
. Apoapsis altitude of 5958 km
= 6378 km + (5858 km + 5958 km)/2

Lageos II's Orbital Period: 223 minutes
(or 13,380 seconds)

Studied Mars' Moons Semi-Axis Values
(From NASA source)

Phobos' Semi-axis 9,378 km

Deimos' Semi-axis 23,459 km

Studied Mars' Moons Orbital Period Values

(From NASA source)

Phobos' Sidereal Orbital Period	0.31891 days (or 27,553.824 seconds)
Deimos' Sidereal Orbital Period	1.26244 days (or 109,074.816 seconds)

Studied Jupiter's Moons Semi-Axis Values
(From NASA source)

Metis' Semi-axis	128,000 km
Adrastea's Semi-axis	129,000 km
Amalthea's Semi-axis	181,400 km
Thebe's Semi-axis	221,900 km
Io's Semi-axis	421,800 km
Europa's Semi-axis	671,100 km
Ganymede's Semi-axis	1,070,400 km
Callisto's Semi-axis	1,882,700 km
Themisto's Semi-axis	7,507,000 km
Leda's Semi-axis	11,165,000 km
Himalia's Semi-axis	11,461,000 km
Ersa's Semi-axis	11,483,000 km
Pandia's Semi-axis	11,525,000 km
Lysithea's Semi-axis	11,717,000 km
Elara's Semi-axis	11,741,000 km
Dia's Semi-axis	12,118,000 km
Carpo's Semi-axis	16,989,000 km
Valetudo's Semi-axis	18,980,000 km
Kallichore's Semi-axis	24,043,000 km
Callirrhoe's Semi-axis	24,102,000 km

S2003J9's Semi-axis	24,234,000 km
Cyllene's Semi-axis	24,349,000 km
Kore's Semi-axis	24,543,000 km
S2003J23's Semi-axis	24,750,000 km

Studied Jupiter's Moons Orbital Period Values

(From NASA source)

Metis' Sidereal Orbital Period	0.294779 days (or 25,468.90 secs)
Adrastea's Sidereal Orbital Period	0.298260 days (or 25,769.664 secs)
Amalthea's Sidereal Orbital Period	0.498179 days (or 43,042.665 secs)
Thebe's Sidereal Orbital Period	0.6745 days (or 58,276.8 secs)
Io's Sidereal Orbital Period	1.769138 days (or 152,853.52 seconds)
Europa's Sidereal Orbital Period	3.551181 days (or 306,822.04 seconds)
Ganymede's Sidereal Orbital Period	7.154553 days (or 618,153.38 seconds)
Callisto's Sidereal Orbital Period	16.689017 days (or 1,441,931.07 secs)
Themisto's Sidereal Orbital Period	130.02 days (or 11,233,728 secs)
Leda's Sidereal Orbital Period	240.92 days (or 20,815,488 secs)
Himalia's Sidereal Orbital Period	250.5662 days (or 21,648,919 secs)
Ersa's Sidereal Orbital Period	252.0 days (or 21,770,000 secs)
Pandia's Sidereal Orbital Period	252.1 days (or 21,781,440 secs)

Lysithea's Sidereal Orbital Period	259.22 days (or 22,396,608 secs)
Elara's Sidereal Orbital Period	259.6528 days (or 22,434,002 secs)
Dia's Sidereal Orbital Period	287.0 days (or 24,800,000 secs)
Carpo's Sidereal Orbital Period	456.1 days (or 39,407,040 secs)
Valetudo's Sidereal Orbital Period	533.3 days (or 46,077,120 secs)
Kallichore's Sidereal Orbital Period (Retrograde)	764.7 days (or 66,070,080 secs)
Callirrhoe's Sidereal Orbital Period (Retrograde)	758.8 days (or 65,560,320 secs)
S2003J9's Sidereal Orbital Period (Retrograde)	766.5 days (or 66,225,600 secs)
Cyllene's Sidereal Orbital Period (Retrograde)	737.8 days (or 63,745,920 secs)
Kore's Sidereal Orbital Period (Retrograde)	779.2 days (or 67,322,880 secs)
S2003J23's Sidereal Orbital Period (Retrograde)	759.7 days (or 65,638,080 secs)

Studied Jupiter's Moons Orbital Resonances

(From Wiki source)

Jupiter's Moons:	Resonances:
Io - Europa - Ganymede	4:2:1

Studied Saturn's Moons Semi-Axis Values

(From NASA source)

Mimas' Semi-axis	185,520 km

Enceladus' Semi-axis	238,020 km
Tethys' Semi-axis	294,660 km
Dione's Semi-axis	377,400 km
Rhea's Semi-axis	527,040 km
Titan's Semi-axis	1,221,870 km
Hyperion's Semi-axis	1,500,930 km
Iapetus' Semi-axis	3,560,850 km

Studied Saturn's Moons Orbital Period Values
(From NASA source)

Mimas' Sidereal Orbital Period	0.94242180 days (or 81,425.24352 secs)
Enceladus' Sidereal Orbital Period	1.370218 days (or 118,386.8352 secs)
Tethys' Sidereal Orbital Period	1.887802 days (or 163,106.0928 secs)
Dione's Sidereal Orbital Period	2.736915 days (or 236,469.456 secs)
Rhea's Sidereal Orbital Period	4.517500 days (or 390,312 secs)
Titan's Sidereal Orbital Period	15.945421 days (or 1,377,684.3744 secs)
Hyperion's Sidereal Orbital Period	21.276609 days (or 1,838,299.0176 secs)
Iapetus' Sidereal Orbital Period	79.330183 days (or 6,854,127.8112 secs)

Studied Saturn's Moons Orbital Resonances
(From source)

Saturn's Moons:	Resonances
Cassini Division – Mimas	2:1

Mimas – Tethys	2:1
Enceladus - Dione	2:1
Titan - Hyperion	4:3

Studied Uranus' Moons Semi-Axis Values
(From NASA source)

Miranda's Semi-axis	129,900 km
Ariel's Semi-axis	190,900 km
Umbriel's Semi-axis	266,000 km
Titania's Semi-axis	436,300 km
Oberon's Semi-axis	583,500 km

Studied Uranus' Moons Orbital Period Values
(From NASA source)

Miranda's Sidereal Orbital Period	1.413479 days (or 122,124.5856 secs)
Ariel's Sidereal Orbital Period	2.520379 days (or 217,760.7456 secs)
Umbriel's Sidereal Orbital Period	4.144176 days (or 358,056.8064 secs)
Titania's Sidereal Orbital Period	8.705867 days (or 752,186.9088 secs)
Oberon's Sidereal Orbital Period	13.463234 days (or 1,163,223.4176 secs)

Final Conclusion

The major part of this book has been to present and explain that:

There exists a set of equation-yielded quantum mechanisms of Kepler's laws.

And that is based on all my potential findings so far.

Equation-Yielded Quantum Mechanism Of Kepler's Second Law:

Kepler's second law posits that

A planet moves in its orbit around the sun, it will sweep out equal areas/surfaces in equal times.

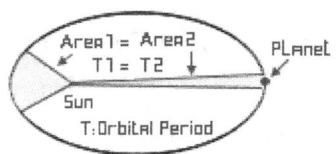

And the potential quantum version of Kepler's second law can be demonstrated by my discovered orbital velocity-radius-period-precession quantizer mechanism through the following equation:

$$T = 2\pi \times ([GM/c^3] \times Dc^3) \qquad (=j2)$$

where:
T is the orbital period of a given captive
c is the speed of light in vacuum
G is Newton's gravitational constant
M is the mass of the captor.
Dc is the orbital quantizer created by the said captor for this captive.

The right-handed side of this quantum equation (j2) contains a constant gravitational energy value "$[GM/c^3] \times Dc^3$" regardless of where the captive is located along its orbit around its captor.

And this constant gravitational energy value alone can explain the quantum mechanism of Kepler's second law. This is because any segment of orbital period of the captive (n x seconds) does not depend on the ever-changing length of its orbital radius nor on its orbital location at all, but instead depends exclusively on a fixed swept surface between the captor and the captive generated by this orbital period segment via the one-second-based gravitational energy value of "$[GM/c^3] \times Dc^3$".

The fixed value of the swept surface between the captor and the captive for a given orbital period segment based on the equation term "$[GM/c^3] \times Dc^3$" from the equation (j2) demonstrates that:

> **There must exist a quantum shuttle that navigates back and forth from the captor to the captive along their connecting quantum axle in order to make Kepler's second law work.**

And

> **By hitting the captor and the captive alternatively and continuously, the said quantum shuttle appears to play the measurement operator that causes the collapse of the wavefunctions of the captor and the captive, hence demonstrating the presence of Heisenberg's uncertainty principle at the cosmic scale.**

By the same token,

> **Any pair of captor and captive behaves like a pair of celestial particles operating as two separate wavefunctions.**
>
> **Each celestial particle must be perpetually in a superposition of two eigenstates: immobile (dead) and mobile (alive).**
>
> **At each captor/captive hit caused by the quantum shuttle – in other words, each time the captor/captive is measured (aka observed) – the wavefunction of the involved celestial particle collapses into a single eigenstate. The new quantum eigenstate (dead or alive) gets translated into the switch to a new classical state of orbital motion (immobile or mobile) of the said celestial particle, hence the change of the latter's momentum or position.**
>
> **This potential quantum mechanism can explain how and why captives classical-mechanically keep orbiting their captor and keep rotating around their axis. This potential quantum mechanism can also explain why and how captors classical-mechanically keep rotating around their axis at the same time.**

Equation-Yielded Quantum Mechanism Of Kepler's Third Law:

Kepler's third law posits that

> The square value of the orbital period of any planet is proportional to the cube value of the semi-major axis of its orbit.

And the potential quantum version of Kepler's third law can be demonstrated by my discovered orbital velocity-radius-period-precession quantizer mechanism through the following equation:

$$T^2 = [4\pi^2 / GM] \times ([GM/c^2] \times Dc^2)^3 \qquad (=j2.6)$$

This equation (j2.6) contain the term ($[GM/c^2] \times Dc^2$) and this term is exactly identical to my previous discovered equation:

$$R = [GM/c^2] \times Dc^2 \qquad (=y2)$$

> where
> R is the orbital radius (aka semi-major axis) of the captive

Therefore,

$$T^2 = [4\pi^2 / GM] \times (R^3) \qquad (=j2.7)$$

This final transitional equation (j2.7) matches Newton's equation of Kepler's third law.

Equation-Yielded Quantum Mechanism Of Perihelion Precession:

My previous finding has revealed that

> The angle (ε) of perihelion precession of any captive can be calculated and predicted by the equivalent two discovered equivalent orbital quantizer-based equations:
>
> $$\varepsilon = (6\pi / Dc^2) / (1-e^2) \qquad (=d1.0)$$
>
> and
>
> $$\varepsilon = (6\pi \times [v/c]^2) / (1-e^2) \qquad (=d0.0)$$
>
> > where:
> > ε is the angle of perihelion precession/shift of any captive
> > Dc is the orbital quantizer of the captive.
> > v is the orbital velocity of the captive
> > c is the speed of light in vacuum

By the same token

> These two equations appear to provide a quantum mechanism alternative to Einstein's spacetime one, know via his equation:
>
> $$\varepsilon = 24\pi^3 \times [a^2/T^2c^2(1-e^2)] \qquad (=e1)$$
>
> > where:

T is the orbital period of the captive's orbital revolution.
a is the semi-major axis (aka orbital radius) of the captive.
c is the speed of light in vacuum
e is the eccentricity of the (elliptical) orbit of the captive

Furthermore,

One of the two discovered equivalent orbital quantizer-based equations (d0.0) of the angle of perihelion precession of any captive contains the Lorentz factor-related β as β = v / c. And this revelation has the potential to provide a quantum link to Einstein's special relativity theory.

Equation-Yielded Quantum Mechanism Of Born Rule:

Born rule is one of the fundamental postulates of quantum mechanics.

Born rule posits that

In a repeated system that comprises some state vector ψ and measures its overlap with another state vector ϕ, the fraction of measurements that find the system in state ϕ is equal to $|\langle\psi|\phi\rangle|^2$ (proportional to the square of the amplitude).

The equation of the orbital velocity-radius-period-precession quantizer mechanism for captives dedicated to the orbital radius:

$$R = [GM/c^2] \times Dc^2 \qquad (=y2)$$

where:
R is the orbital radius of a specific captive.
M is the mass of the captor.
Dc is the orbital quantizer created by the captor for a specific captive.
G is Newton's gravitational constant
c is the speed of light in vacuum

appear to obey the Born rule,

should the quantum state vector ψ here be "GM/c^2", which is the gravitational energy quanta.

Two Speeds Of Gravity At Quantum Level Revealed By Equations Of Orbital Velocity-Radius-Period-Precession Mechanism:

Based on my findings so far,

Gravity at the quantum level appears to operate at two speeds whose values are the square and the cube of the speed of light (denoted respectively as "c^2" and "c^3"). These speeds can be categorized as one of the quantum speeds of gravity.

When and how Gravity operates at the cube of the speed of light ("c^3"):

The equivalent quantum orbital period equation (j2) of Kepler's second law

$$T = 2\pi \times ([GM/c^3] \times Dc^3) \qquad (=j2)$$

reveals that

Gravity appears to operate at a speed whose value is the cube of the speed of light (denoted as "c^3") to generate and maintain the orbital period of captives.

This finding is due to the fact that

The fixed value of the swept surface between the captor and the captive for one second-based orbital period, according to the term "$[GM/c^3] \times Dc^3$" of the equation (j2), can only be possible if and only if

The orbital period dedicated quantum shuttle (measurement operator) is able to travel from the captor to the captive then return during one second at the speed of the cube value of the speed of light (c^3).

The orbital period dedicated quantum shuttle achieves the said travel by spending all its specific gravitational energy quanta, defined by the equation term "$[GM/c^3] \times Dc^3$".

When and how Gravity operates at the square of the speed of light ("c^2"):

The orbital period quantizer mechanism equation (y2) - equivalent to classical Kepler's second law -

$$R = [GM/c^2] \times Dc^2 \qquad (=y2)$$

reveals that

Gravity appears to also operate at a slower speed whose value is the square of the speed of light (denoted as "c^2") to generate and maintain the orbital radius of captives.

This finding is due to the fact that

The fixed value of the length between the captor and the captive maintained during each second, according to the term "$[GM/c^2] \times Dc^2$" of the equation (y2), can only be possible if and only if

> The orbital radius dedicated quantum shuttle (measurement operator) is able this time to travel from the captor to the captive then return during one second at the speed of the square value of the speed of light (c^2).

> The orbital radius dedicated quantum shuttle achieves the said travel by spending all its specific gravitational energy quanta, defined by the equation term "$[GM/c^2] \times Dc^2$".

Should these two gravity's quantum speeds exist, there must exist either

> **Two types of particle whose one can move at the cube of the speed of light (c^3) and the other move at the square of the speed of light (c^2).**

or

> **One type of particle that has a switch to allow it to move either at the cube of the speed of light (c^3) or at the square of the speed of light (c^2).**

These types of particles must be the engine of the said quantum shuttles (measurement operators) that travel back and forth between the captor to the captive.

Orbital Velocity-Radius-Period-Precession Quantizer Mechanism:

Based on all my findings so far,

> The quantum equation (j2) of Kepler's second law does not happen in a vacuum, but is just part of a set of equations that can be summarized as the
>
> > Orbital Velocity-Radius-Period-Precession Quantizer Mechanism.
>
> And all these equations share the same core quantum element called an "Orbital quantizer".
>
> As the core element of the orbital velocity-radius-period-precession quantizer mechanism, the orbital quantizer is defined by the following equation:
>
> $$Dc = c / V \qquad (=s1)$$
>
> > where:
> > V is the actual orbital velocity of the captive (present in Newton's equation of orbital velocity)
> > c is the speed-of-light in vacuum

The equation (s1) yields the orbital velocity of a given captive, as an integral part of the orbital velocity-radius-period-precession quantizer mechanism.

The same orbital quantizer determines also the orbital radius of a given captive via the following equation:

$$R = [GM/c^2] \times Dc^2 \qquad (=y2)$$

where:

R is the orbital radius for a specific captive
Dc is the orbital quantizer for this captive
M is the mass of the captor
G is Newton's gravitational constant

GM/c^2 is the quantum core radius for the captive

The equation (y2), also as an integral part of the orbital velocity-radius-period-precession quantizer mechanism, yields the orbital radius of a given captive.

Because of the constant value of "$[GM/c^2] \times Dc^2$" inside the equation (y2), the mean orbital radius of a given captive must remain constant for every orbital cycle.

And the constancy of the said mean orbital radius of a given captive can explain the elliptical orbit nature of captives described by Kepler's first law.

Concretely, during every orbital cycle, the captive may be gradually pushed away from the captor or gradually pulled towards it, due to its intrinsic momentum, but overall, the captive always manages to restore its orbital location designated by the orbital velocity-radius-period-precession quantizer mechanism via the equation (y2), hence the existence of the mean orbital radius.

Orbital Velocity Of Captives As Function Of Orbital Quantizer Than That Of Mass Of Captor:

The orbital velocity-radius-period-precession quantizer mechanism has revealed that

The orbital quantizer is the unique determining factor of the three core gravitational features of any captive: orbital radius, orbital velocity and orbital period.

As far as the orbital velocity is concerned,

The equation

$V = c / Dc$ (=s5)

which comes from   (=s1)

$Dc = c / V$ (=s1)

shows that the orbital velocity is a function of a frequency that depends solely on the speed of light (c), not on the gravitational mass "GM" as defined by Newton's law of orbital velocity of captive "$V^2 = GM/R$".

At the same time, the equations that determine the orbital radius (y2) and the orbital period (j2) depend both on the speed of light (c) and on the gravitational mass "GM", as shown below:

$T = 2\pi \times ([GM/c^3] \times Dc^3)$ (=j2)

$R = [GM/c^2] \times Dc^2$ (=y2)

One can notice that the two equations above (y2) and (j2) share the same value of gravitational mass "GM". It turns out that "GM" can be associated with a multiplier different than 1 because this gravitational mass multiplier mechanism has already been discovered to exist in different gravitational equations such as Schwarzschild radius ($2GM/c^2$), Einstein's time dilation ($2GM/c^2$), Einstein's gravitational lensing ($4GM/c^2$), Einstein's gravitational redshift ($4GM/c^2$), captives' perihelion precession angle ($2\pi \times 3GM/c^2$).

By the same token, it is conceivable that two or more captives with different orbital radii can host the same orbital quantizer (Dc), hence the same orbital velocity (V) - as $V = c / Dc$ according to (s5) - so long that Newton's gravitational constant G of each captive has a different multiplier. And the mechanism of different multipliers of G shows that there is no need of multiplying the actual mass of the captor as Newton's law equation "$V^2 = GM/R$" requires in the case of the flat galactic rotation curve anomaly.

Edition Changes Notice

The changes between editions up to this one are presented hereafter:

Changes from Edition 1 to 1.1:

Fixing unit errors of $GMsun/c^3$, $GMearth/c^3$, $GMmars/c^3$, $Gmjupiter/c^3$, $GMsaturn/c^3$, $GMuranus/c^3$, ...

That should be seconds not meters

Adding Definition of Nature of "GM/c^3" Unit as

Quantum observer-measurer-communicator energy lifespan